BEYOND THE

KNOWN UNIVERSE

Also by I. M. Levitt

I. M. Levitt

Director Emeritus, The Fels Planetarium of
The Franklin Institute

BEYOND THE
KNOWN UNIVERSE

From Dwarf Stars

to Quasars

The Viking Press   New York

To Alice—with love

Contents

Illustrations follow pages 20, 52, and 84.

Introduction

You are in the living room and reading the evening paper. Within the wink of an eyelid, the room, the house, the entire earth have been turned into a cinder that will eventually become a hot gas, with a temperature which may rise to hundreds of millions of degrees—scores of times hotter than the core of the sun. An observer from afar would see this part of space instantly become luminous, then brighter, and then, depending upon the distance, perhaps brighter than anything in the sky, including his own sun. Does this sound bizarre—weird? Yes, it does; and yet what we are discussing is the sudden occurrence in our part of the universe of the phenomenon of collapse, disappearance, and, perhaps, reappearance of stars—events which astronomers may now be observing in distant parts of the universe.

This is a dramatic "happening," yet it is a phenomenon which unfolds from the application of the laws of physics as we understand them today. We are simply indicating the reappearance elsewhere of a star which has collapsed to an extraordinarily high density, with zero dimension, into a "singularity," and moved out of our universe. The saying, "truth is stranger than fiction" certainly applies, for no Jules Verne or H. G. Wells could have possibly conceived a situation nearly as incomprehensible. We are discussing a singularity—one of the paradoxes of this violent universe—and this, in essence, is the plan of the book. It plans to explore those exotic celestial objects which behave in bizarre and unfamiliar ways —objects which stretch the imagination of the scientist beyond reasonable limits.

The twentieth century has brought into being an exciting science

and a technology which have projected the questing mind of man to
the far reaches of the universe, indeed beyond the known universe. His
technology has expanded the real and mental horizons until the mind
is overwhelmed as it tries to free itself of the shackles of terrestrial
prejudices. The application of many disciplines, using the laws of
physics to explain exotic and intriguing objects, uncovered by the new
science, has shown that this remarkable universe in which we live is,
for the most part, unknown to us. In those areas in which knowledge
becomes available, we find that not even the most audacious mind is
prepared to accept this knowledge in the form in which it is presented.
And, if we are staggered by the awesome picture presented by these
paradoxes, let us recognize that no other science in the history of man
has achieved such a phenomenal increase in basic knowledge as the
science involving these unique objects. Their unveiling has all taken
place in the past decade. The insatiable, imaginative hunger for fact
and reality has permeated astrophysics to bring to light these provocative celestial bodies.

Essentially, this book is a story of the mind-stretching activities
in which scientists have been absorbed for the past several decades.
Definitive answers to the questions posed by these discoveries are not
available, and it may be that generations as yet unborn will explore
them for possible answers. However, at this time, scientists gifted
with insights and consummate skills have provided a sizable body of
knowledge about these rarities and this is what this essay is all about.

I am delighted to acknowledge the aid I have received from
a select group of astronomers and physicists who have applied their
ingenuity and superb skills to the problems with which these writings
are concerned. Their profound and brilliant reasoning has fashioned
these problems into a fascinating story that I hope to capture. First,
I would like to express my appreciation to the late Dr. Roy K.
Marshall, formerly director of several planetariums and my prede-
cessor at the Fels Planetarium. His uncanny ability to make ab-

struse concepts intelligible to the layman has influenced me over the years, and I am certain his suggestions have gone far to make this work appeal to the nontechnical reader. To Doctors Louis Green of Haverford College; Malvin A. Ruderman of Columbia University; Remo Ruffini of Princeton University and the Institute for Advanced Study; A. G. W. Cameron of Yeshiva University; Freeman J. Dyson of the Institute for Advanced Study, Princeton; Donald H. Menzel of Harvard University; and Jeremiah P. Ostriker of Princeton University—I owe a debt of gratitude for their gracious review of sections of this manuscript which fell within their special spheres of competence. If, in adapting their researches and writings, I may have unintentionally warped their meanings and intent, I accept full responsibility for any errors which may have been introduced.

I owe a special debt of gratitude to John Gorsuch, who, for over fifteen years, has illustrated my space column. His special talent in portraying difficult concepts to make them understandable to the layman is a unique and magnificent gift. I am certain that these superb illustrations will do much to enhance the value of this book.

Grateful acknowledgment is given to the Kitt Peak National Observatory and the Mount Wilson and Palomar Observatories for permission to reproduce their photographs.

I would also acknowledge my debt to my long-time secretary, Margaret C. Ellson, for her invaluable assistance in the preparation of the manuscript.

BEYOND THE
KNOWN UNIVERSE

1

Of Particles, Radiations,

and the Universe

Concentrated efforts over the past decade have been devoted to outer space and a veritable harvest of significant results has cascaded down upon the contemporary scene. But we should also be reminded that there is inner space. And by inner space I do not mean the earth or any portion of it. By inner space I mean the domain of those particles that make up every bit of stuff in us, in the earth, indeed in the entire universe. To undertake a detailed discussion of the largest and most violent members of the universe, we must first acknowledge the vital role played by the smallest. It may not be entirely fortuitous that knowledge of the various submicroscopic particles that comprise the world around us contain within themselves the information and background which could lead to an understanding of the largest, most distant, and most energetic members of the universe. But, unfortunately, our knowledge of these submicroscopic particles is shrouded in an impenetrable haze and thus severely limited. It is hoped that new knowledge of their presence and behavior will be revealed in the future to fill in crucial gaps. This new reach for knowledge will continue, for scientists have an intuitive feeling that other particles exist which may play a dominant role in the behavior of matter in the universe. Unfortunately, these new particles are so minute as to remain permanently in a realm where physical exploration is impossible. The analytical mind of the scientist has projected itself into this almost forbidden realm and has deduced answers that permit him to predict reactions and behaviors which represent the sole proof that his speculations have been right or at least partially correct.

That the known universe is composed of particles of great variety is not a new concept. More than 2000 years ago Democritus, a Greek philosopher, had advanced an atomic theory in which he speculated that all matter was composed of minute, invisible, identical particles. By varying the number of particles in a substance, its nature varied. These particles, atoms (from the Greek *tomos*— cut, and *a*—a negative; therefore atom means uncuttable, or indivisible) formed the dominant theme of the concepts Democritus created, but little did he realize how close he approached the truth. And it should be noted that in the modern sense, Democritus really had molecules in mind and not atoms; nevertheless this theory constituted a major conceptual advance.

If, in our discussion of atoms, we pursue history, we must take a major leap into the nineteenth century. In the first decade of that century John Dalton presented the opinion that there was really a limited variety of elementary particles, each of them fundamentally different in some way from all the others. These were the elements— such as hydrogen, oxygen, carbon, gold, and more than a hundred others. It was believed that these elements could not be broken up or dissociated into other substances. Today, knowledge of these elements provides a completely different picture as to their behavior. The work of Dalton was singularly perceptive, for it postulated a minimum number of particles from which all matter in the universe is fashioned.

We now move ahead, almost a century, to 1896. In that year, the French physicist Antoine Henri Becquerel made the startling discovery that a piece of potassium uranyl sulfate emitted radiations which had penetrated the wrappings on a photographic plate and had impressed the silhouette of a key on the plate as though the plate had been exposed to the image of the key. By experimenting he discovered that any chemical containing uranium behaved in the same way. The significant point about this discovery was that some elements were radioactive; that is, they spontaneously decayed into some other material which conclusively indicated that atoms were not impregnable but possessed structure which in the case of the radioactive elements would break down to less massive elements by the emission of radiations or particles from the nucleus.

The concept of particles—positive as the proton, and negative as the electron—was known at the turn of the twentieth century. In his lectures, Samuel A. Goudsmit had this to say about electrons and protons: in the case of the electron, blow up a bubble. Then dip a small brush into a little pot containing charges of negative electricity. Pick up a single unit, or charge, of negative electricity, and paint the bubble with the brush to cover it with the single unit of negative charge. Now take a pin and puncture the bubble. What you have left is a spherical distribution of negative electricity, with its electric field surrounding it. This he called an electron.

We speak of a negative charge, but what do we mean by this? What is the magnitude of the charge? On my desk is a lamp with a 100-watt bulb. If 10 million trillion* electrons pass through the wire every second, that lamp will light up. Thus the charge on each electron is minute. Being so small, electrons cannot weigh very much. If we placed a single drop of water on one pan of a most delicate balance, or pair of scales, we would have to put 100 trillion trillion electrons on the other just to balance it.

For a proton you do the same thing. But now you blow up another bubble, and you dip your brush into a pot containing charges of positive electricity and pick up one positive charge. Now paint

* In this book the American notation is used, in which a billion is a thousand million, and a trillion is a million million.

the bubble with the single electrical charge and the paint will be thicker and therefore it will have a stronger field. Now prick the bubble so that it disappears. What we have left is the proton.

Even after knowledge of the electron and proton particles became familiar, scientists were still asking highly pertinent questions. They could not understand some aspects of atomic theory and another new particle was needed. While it had mass, the particle could not have an electrical charge. In 1932, James Chadwick discovered the neutron, and the picture of the core of the atom became understandable.

All elements had nuclei composed of protons and neutrons with electron shells circling these nuclei. The factor which distinguished one element from another was the number of protons in the nucleus. In hydrogen there was but one proton. Thus, every atom which contained a single proton was a hydrogen atom. If the nucleus carried two protons, the element had to be helium. If there were four protons, the element had to be beryllium. And if there were eight protons in the nucleus, the element was oxygen. The number of neutrons varied with given numbers of protons. As an example, a single proton in the nucleus meant a hydrogen atom. But, if there were a neutron associated with the proton, the atom would still be hydrogen —but heavy hydrogen, or deuterium—for its mass was increased by 100 per cent. If the nucleus contained one proton but two neutrons, then again we had hydrogen but, this time, it was heavy-heavy hydrogen, or tritium. These other forms of hydrogen were called isotopes.

The neutron is a neutral particle, one without an electrical charge. What are its physical characteristics? The answer is that a proton and an electron combined to create this neutron.

Thus, in 1932, we could speak of three fundamental particles of which all matter was composed. However, the end was not in sight, for in that same year another particle which had been predicted was discovered—the positive electron, or positron. The positron possessed the same mass as the electron but had an opposite electrical charge: it was positively charged. Now there existed four fundamental particles.

There is still another particle. We are referring to the photon—a particle of light. Light behaving like a particle of zero mass, moves with a speed of about 186,000 miles per second. Light also behaves as a wave—in point of fact, Sir Arthur S. Eddington preferred to think of light as a "wavicle" so that its behavior could be described in either of these two modes. The only point we might stress at this time is that photons possess an entire spectrum of energies. If a photon has extremely high energy we call it a gamma ray and as the energy content decreases we go to X rays, ultraviolet radiation, visible light, and then into the infrared part of the electromagnetic spectrum. All of these are photons and all move as light with the speed of light.

In studying radioactive decay, scientists discovered that electrons, or "beta" particles, were being emitted by the nucleus of the atom. This told them that the nucleus could create and eject electrons. They also observed that when a beta particle appeared, the element appeared to undergo a change in character. The neutron was being transformed into a proton by the emission of the electron, and the element changed because of the increase in the number of protons. Atoms also participated in other reactions that permitted them to add neutrons. With additional neutrons, the element remains the same (the proton number is preserved) but the element becomes more massive because of the added neutron, and a different isotope of the element is created.

For the moment, let's return to the breakdown of the neutron (in a nucleus) into a proton by the emission of an electron. This reaction was accepted for a long time with no questions asked; the physical equations appeared to balance. The trick word in this sentence is "appeared." Actually, the equations did not balance out. When further experiments were performed, and the results carefully assessed, it was discovered that the velocities of the emitted electrons were too small in some cases. This indicated to scientists that some energy was missing. As long ago as 1930, the physicist Wolfgang Pauli suggested that perhaps a completely new and strange type of particle was also emitted simultaneously with the emission of the electron. Enrico Fermi developed the idea and called it a neutrino, which in Italian means "neutral little one." What accompanies the electron is an antineutrino; we will discuss the "anti" character in detail shortly. The neutrino and electron share the energy lost by the nucleus in all proportions—and now the equation balanced. However, it was discovered that the electron receives all of the available mass and the neutrino receives none. This meant that the mass of the neutrino is zero but that as it always moves with the velocity of light, it does possess energy. The neutrino is thus a particle of zero mass but with a significant amount of energy moving at 186,000 miles per second. Neutrinos are really ghost particles—cosmic ghosts. We cannot see them—we can barely detect them—but we know they must exist.

Now, the question was posed: how is the neutrino different from the photon, which also has zero mass and moves with the speed of light? The answer is in the spin of the particle—the photon has twice the spin of the neutrino.

In what follows, as we delve into the character of the paradoxes of the universe, we will discover that the neutrino is a most important particle. It is the mechanism by which stars can radiate a significant portion of their energy from deep in their interior. For this reason, it will not be out of place to detail its character. Interaction of the neutrino with other matter is an extremely rare occurrence. To appreciate the rarity of this event, we might point out that a very high-energy photon can penetrate solid lead for a few centimeters and take about 10 billionths of a second doing so before interacting with the lead. A neutrino of the same energy can travel for about fifty years before interacting. To stop the neutrino, the lead shield would have to be thicker than the distance between the earth and the star Arcturus. Despite their rare interactions, neutrinos are still numerous. They are generated deep in the interior of the stars; in the case of the earth we are bombarded by neutrinos from the sun with perhaps 100,000 passing through our bodies in the time it takes to read this sentence. They can pass through the earth, the sun, and even the less massive stars without being stopped; it is estimated that on earth we have one interaction for every trillion neutrinos which pass through it.

When an atomic nucleus ejects a positron, a neutrino is also emitted to conserve the angular momentum. The antineutrino is a neutrino with opposite electrical characteristics. It is an anti-particle. Just as we have antineutrinos and electron-positron particles, so are also found antiprotons and antineutrons. Apparently, there is a symmetry in the particles that are known to modern physics so one can say with some certainty that the presence of one particle indicates that its opposite, or antiparticle, is also present. The antiparticles have been called antimatter, and on a rare occasion a particle and an antiparticle will collide. When they do, they annihilate each other with the production of energy in the form of photons of various energies. Thus, an electron and a positron can collide to give rise to two gamma rays, which are simply very high-energy photons. Occasionally, the reverse process, called "materialization of light," will occur and a gamma ray will give rise to an electron and a positron.

We have still not exhausted the particles which scientists have discovered. It has been estimated that there are in excess of 1000 particles which make up all matter in the universe. However, our objective is not the study of particles. We are concerned only with those particles whose presence is of major importance in stellar interiors, in order to discover what reactions take place there. For this reason, we will restrict our attention to only two other classes of particles. The first are the mesons, of which there are several types and which come into existence when a proton and antiproton annihilate one another; the second are the heavy baryons, or hyperons, whose mass is greater than that of protons or neutrons.

These, then, are the particles with which we will deal. Now let's explore some of the physical characteristics of the particles. Most of the mass of the atom is contained in the nucleus. If we assume that the electron has a mass of one unit, then a proton has a mass 1836.1 times that of the electron. As the neutron has the characteristics of the combined electron and proton, its mass is not very different from that of the proton. From this, it is readily seen that most of the mass of an atom resides in the nucleus. Its volume is about 1 trillionth that of the volume of the atom. If the nucleus were magnified until it were the size of the Empire State Building, the atom would reach out one sixth of the way to the moon. Thus, we find that as in the solar system, the atom is mostly empty space, and when an electron is removed from the hydrogen atom—to make the atom ionized—then the mass of the atom is the mass of the nucleus and the size of the atom is the size of the nucleus. This means that Nature can pack a tremendous mass into a small volume. If we had a gas composed of ionized hydrogen and permitted the nuclei to come as close as possible, then we would essentially have a nuclear fluid with a density 2300 trillion times the density of water. When we begin detailing the core of the neutron stars, this is the density we will encounter.

Having introduced the particles of which the universe is composed, let's examine the way in which astronomers interpret messages which ride on the light of the stars to the earth. To begin, our only clue to the universe is by some form of radiation. Whereas other scientists in their studies and investigations can touch, taste, smell, hear, and see the results of their experiments, the astronomer is severely handicapped because he can use only one sense—that of sight—in the pursuit of his studies. Yet with this single sense he has reached out literally to the edge of the space to fill in the gaps in the story of the universe. The achievements of the astronomer with this significantly curtailed armamentarium are truly remarkable.

Electromagnetic radiation is a ubiquitous phenomenon. Everything in the universe emits radiations of one form or another unless it happens to have a temperature of absolute zero. Except for this single (and literally impossible) condition, we can expect electromagnetic radiations from all material bodies in the universe.

For this reason, as you gaze at someone, you are being bombarded by reflected radiations and by radiations which originate in and emanate from that someone. Even though you are unaware that there is anyone in a completely darkened room, that person can be detected by using the sniperscope, or other appropriate type of sensing instrument. In the case of a human being with a body temperature of 98.6 degrees F., the radiation is in the infrared part of the spectrum. A blind person, given a suitable sensing instrument to detect infrared radiation, can sense the positions, distances, and number of people in a darkened room even though he cannot see.

As one looks into the sky one sees the stars, the planets, the sun, and the moon. All these celestial bodies emit radiations of various types. The stars, because of their high temperatures, emit copious amounts of radiations over a broad region of the electromagnetic spectrum. Unfortunately, we, on the surface of the earth, reside at the bottom of a hazy, turbulent, shimmering sea of air which distorts, alters, and filters these radiations. Thus, the complete, true picture of what is pouring in from the stars is not perceivable from the surface of the earth. However, the new astronomical observatories orbiting the earth, high above the sensible atmosphere, can detect the entire spectrum of radiations arriving from these bodies. In the future, observatories on the airless moon will pursue the same function.

Broad-band radiations do not tell us too much except that a vast amount of energy is being emitted by the distant stars. To determine the detailed characteristics of the celestial bodies, we must break up the light of the star into its spectrum and read the message in the resulting spectrum. The spectrum can disclose much of the nature of the distant celestial object. These astronomical objects emit various kinds of spectra, and by interpreting the spectra we learn what is taking place on and in the star.

To unravel the message of light, we return to basic principles. The sources of radiation are the interactions of the atoms and molecules in the celestial bodies. We will ignore molecular radiations and concentrate on the atom. It is the ability of the atoms to absorb and emit energy under certain specific conditions which provides the clues needed to interpret what is seen.

To begin, we might consider the hydrogen atom. This is the most uncomplicated of all the atoms, for its nucleus is the proton, and orbiting this proton is one electron. If the hydrogen atom encounters a "packet," or "pulse," of energy, such as a photon, the hydrogen might absorb this energy with the result that the electron will move into a higher orbit as it circles the proton. After a tiny fraction of a second, the electron may drop back to its previous position and, in doing so, it must emit energy of precisely the same magnitude as it absorbed. The magnitude of the energy absorbed or emitted is precisely known and shows up as a line in a spectrum. Thus, when one examines the spectrum of an object and finds lines of particular wave lengths it is possible to identify the element responsible for the lines. Just as no two human beings possess the same fingerprints, so no two atoms generate the same lines, either in absorption or emission. If we know the wave length of the energy emitted, we know where to look to find the line. This is true of all the atoms in the periodic table. Once we have obtained a spectrogram and can measure the positions of the lines, it is possible to identify the elements.

The physical laws governing the type of spectrum emitted by various celestial objects have been known for a long time, and a simple experiment can illustrate their operation. Suppose one is holding a glowing lamp bulb that contains a bare filament. The filament will be seen to glow brightly. Now suppose one takes a spectroscope, the heart of which is a glorified chandelier crystal, and permits the light to pass through it. The spectroscope will emit a rainbow of color, with red at one end and blue at the other. The band of color will be continuous—no breaks will be seen. In fact, this is why it is called a continuous spectrum. All glowing solids, liquids, or highly compressed gases emit continuous spectra.

Picture a glass tube containing a gas. It can be hydrogen or helium or any atomic gas. Because we are more familiar with neon, let's suppose this tube after having been evacuated has had a pinch of neon gas inserted into it. If an electric discharge is passed through this gas, it will begin to glow with the characteristic red color of the gas. Now when we pass this light through the spectroscope, we can see a pattern of bright lines of various colors—the emission lines for this gas. These are produced as the electrical discharge makes the electrons of the atom jump to larger orbits, from which they then fall to lower orbits, emitting the energy in discrete parts of the electromagnetic spectrum. The positions of these lines are precisely known for all gases, and in this fashion, if at first we have no idea what the gas is, we can determine its identity by measuring the positions of the lines.

Let's go back to the electric light bulb. Suppose a cloud of gas at low pressure is introduced between the light and the spectrograph. Before the gas was introduced we saw a continuous spectrum. Now the continuous spectrum is crossed by dark lines. The atoms of the gas will steal a little bit of light from various parts of the spectrum and this shows up as dark lines. If we measure the positions of the dark lines, we can again determine the identity of the cool gas that was introduced between the filament and the spectroscope.

These are the three types of spectra the astronomer meets in his studies of celestial objects. While the circumstances in the sky may be totally different from what scientists do in the laboratory, the basic mechanism for the production of the various types of spectra holds for the entire universe.

In our discussion of hydrogen it was indicated that when we stripped the electron from the hydrogen atom and only the core, or nucleus, remained, we said the hydrogen atom was ionized. In some cases, the atom will absorb only so much energy to sunder the atom into a positive and negative component. The electron with a negative charge has escaped, and as the atom started off neutral it means the proton has a residual positive charge. Thus, ionization can create charged particles that are negative and positive.

The spectroscope can tell us how hot the star surfaces are. While qualitatively the spectrum of one glowing solid is identical with that of another glowing solid, quantitatively there is a difference, depending upon the temperature. A piece of iron barely hot to the touch is not glowing yet it is radiating energy—energy of very long wave length, not visible to the eye, but susceptible to the hand as heat. As the temperature of the iron is increased, the glow begins, at first a dull red, then a bright red, then orange, then yellow, then white. If the spectroscope is used during this process it can be seen that at the start the red is the strongest part of the spectrum, although the other colors are present. As the temperature increases, the intensity of each color increases—but not at the same rate. At all times all colors are represented, but the relative strengths of the various colors determine what the mixture will look like without the spectroscope.

By comparing the intensities of the various colors through the spectrum, the temperature of a star may be deduced. This is, of course, the effective radiating temperature of the star, not that of a point deep inside. For the sun the temperature is about 5700 degrees (all temperatures mentioned are on the absolute scale called Kelvin–K°). The cool red stars have surface temperatures of about 2500 degrees, while the very hot stars have temperatures of 50,000 degrees, or more.

The lines in the spectrum may be shifted toward the red or blue end of the spectrum to provide scientists with a valuable tool in interpreting line-of-sight motions in space. This Doppler-Fizeau shift, or what is called the Doppler shift, may best be explained by imagining a train racing toward you with its horn wide open. As the train approaches, the sound waves pile up and the result is that you hear a higher pitch. As the train passes and moves away from the observer, the speed of the sound waves is diminished by the speed of the train. The sound waves now seem to be strung out, and appear to be at a lower frequency, or pitch, as the sound reaches the ear. The Doppler shift can then provide an index to the line-of-sight motions of celestial objects. It is with this method that the velocity of approach or recession is acquired. It is this mode of operation which will be used in detail to account for some curious and, at times, incredibly fast motions of the celestial paradoxes.

2

Distances,

Brightness, Energy—

the Keys to Knowledge

Our knowledge of the known universe is inexorably linked with our ability to determine distances in space. From time immemorial the question "how far?" has assumed a vital role, one which has hindered the astronomer in his quest for knowledge of the character of the universe in which he lives. However, no matter how much man willed this knowledge, it could not be obtained until delicate and superbly designed instruments became available to him. Thus, through the centuries, as knowledge of the physical world advanced, the curtains which masked the measuring rods of space remained intact. For literally centuries the philosopher and astronomer pondered these distances and diligently sought means of measuring them. But all was in vain, for the necessary tools could not be fashioned. Finally, when the telescope had been in use for many years and early geniuses had applied the findings of the telescope to their intellectual gifts, the stage was set for a marriage of precision mechanisms to the exquisite optics of the telescope to fashion an instrument which could resolve the problem of distances. When the shackles were sundered, many astronomers pooled their knowledge, skills, and insights to reveal the colossal distances at which the star systems lie.

In 1838, three astronomers (in different parts of the world) successfully measured the distance to a star. Friedrich Wilhelm Bessel, in Germany, measured the distance to the star 61 Cygni. Russia's ranking astronomer, Friedrich von Struve, uncovered the distance to the magnificently bright star of the summer sky—Vega. At the Cape of Good Hope, in South Africa, Thomas Henderson measured the distance to what was to be acknowledged as the nearest star to the sun—Alpha Centauri. In each case, the astronomers measured an inconceivably small angle to determine what the astronomer calls the parallax. The common ground for the discovery of these distances is that the objects of their study—these stars—were relatively close to the earth.

There is an ancient story told of the astronomer who, following a lecture, was approached by a sweet, dear little old lady who said, "You know, I understand exactly how you determine how far away the stars are, how big they are, how hot they are, and how fast they move but, for the life of me, I cannot understand how you astronomers discover their names." I would assume that the little old lady also did not know how the astronomers determined these other characteristics of the stars.

The distances to the stars, while involving measurements of the most precise type, are determined by a relatively simple problem in trigonometry. It operates on a process which depends on the principle of parallax—the apparent change of position of an object as the observer's position changes. More precisely, we should say it is the apparent change in direction toward a body as the observer's place changes, for only the direction appears to change.

In the Fels Planetarium, of The Franklin Institute, we often encourage our audiences to participate in an exercise to demonstrate the principle of parallax. We ask the visitor to hold up a finger in front of his face. We tell him to look at it with one eye—the right eye. Inevitably, he will find a star against which the finger will position itself. Then, without moving the finger, we suggest he close the right eye and open the left. In doing this, he will note the position of the finger against the background of the stars. The finger will appear to have changed its position. You can try this as you read this book. Hold your finger in front of your face and blink first one eye and then the other; the finger will appear to jump against the background of the wall. The change in its position is parallax, and we are seeing a parallactic displacement.

This experiment can be refined by holding the finger close to your face and blinking your eyes and then holding the finger at arm's length and repeating the blinking. The farther the finger is from your face, the less it appears to jump against the background of the wall.

The surveyor uses this principle to measure the distance across a river without crossing it. He measures carefully the distance between two points on his side of the river and then picks out a tree on the river bank. By setting up his transit or angle-measuring instrument on both points, he measures the angle between the tree and the other point. This uniquely defines a triangle, for he has two angles and the included line, which is all the information he needs to determine the width of the river. An extension of this concept to the measurement of the distances to the stars is simple.

The earth revolves around the sun once a year in an orbit which is 186 million miles in diameter. At one time of the year, say January 1, we are at one point in the earth's orbit. Six months later, we are 186 million miles away from that point. (This statement is not quite true, for the sun also moves, but we will ignore this motion in this explanation because it is so minute.) If we observe a star from these two positions, we have again uniquely determined another triangle whose baseline is the diameter of the earth's orbit and the angles have been measured at the earth's position for January 1 and July 1. While we have uniquely determined the triangle, the solution is not without problems; the angles which are measured are incredibly minute and thus must be measured with the utmost precision. These angles are determined by the shift of the target star against the very faint background stars which, on the average, will be much more distant than the target star. The angle is less than 1 second of arc for the nearest star, which is about 25 trillion miles away.

What is 1 second of arc? Look into the sky at the moon. The moon is about 30 minutes of arc in diameter. Divide the moon into 30 slivers and each sliver is then 1 minute of arc. Now take a sliver and divide it into 60 parts. The width of this exceedingly narrow 1/60 part is 1 second of arc and the angle measured to the nearest star is approximately 3/4 of this. The other stars are much more distant and the angle is considerably smaller. What is 3/4 second of arc? Let the circumference of a penny represent the orbit of the earth, i.e., 186 million miles in diameter. On this reduced scale, the nearest star would be 6.472 miles away.

It is most difficult for the average layman to carry large numbers in his head and the astronomer is no different; he has devised a type of shorthand to eliminate cumbersome numbers. He uses the speed of light upon which to build another and more easily understood unit of distance. Light travels with a speed of 186,000 miles a second. In one year, light will travel about 5800 billion miles. This is what the astronomer calls a light-year. The light-year has become one of the basic units of astronomical distance measurement. In these terms, the distance to the nearest star, Alpha Centauri, is 4.3 light-years.

Because the angles being measured are so tiny, the angles for the more distant stars become so minute that astronomers no longer consider their determinations valid, for the measurement errors are of the same magnitude as the distances. When this happens, we must resort to some other method to determine distances. However, before the trigonometric parallax system breaks down, the astronomer has derived the distances of over 6000 stars. How does the astronomer make a determination beyond these distances? This difficult problem was tentatively resolved when astronomers discovered a certain type of variable star that varied in brightness with a predictable regularity. The reason the word tentatively is used is because no cepheid variable has been discovered whose distance can be determined directly by the trigonometric parallax method. Unfortunately, at this time there is no overlap in these two methods.

These cepheids were found to vary in a way that permitted the astronomer to determine their average luminosities. Thus, a star which varied with a period of a half day was 40 times brighter than the sun, while one which varied with a period of 40 days was 3000 times brighter than the sun. To those who wonder how a start was made on this distance-determination scheme, we might point out that in the Magellanic Clouds—satellite galaxies of the Milky Way, which are seen from southern latitudes—were found many cepheid variables. As the distance to all the stars in these clouds were approximately the same, the period-luminosity curve could be derived for them.

While these luminosities were not precise because the cepheids were at varying distances in the Magellanic Clouds and there were some local, absorbing materials which could dim the light of the stars, the luminosities could not be off by more than a small fraction of a magnitude. But these luminosities or absolute magnitudes still did not provide a distance to Magellanic Clouds and a calibration for the cepheid variables. In 1952, the absolute magnitudes of the cepheid variables in the great galaxy in Andromeda were determined. The apparent magnitudes could be measured from a photographic plate. When the absolute and apparent magnitudes were available, then the application of a simple formula derived the distance. Once the distance had been checked out for several galaxies, then the distances to any cepheid variable of measurable period could be determined and these stars became excellent celestial measuring rods.

There are no gravitational forces acting on the inside of a large earth-orbiting box. The density of a gas at sea-level pressure is very low.

The 300-foot orbiting box, filled with Ping-pong balls.

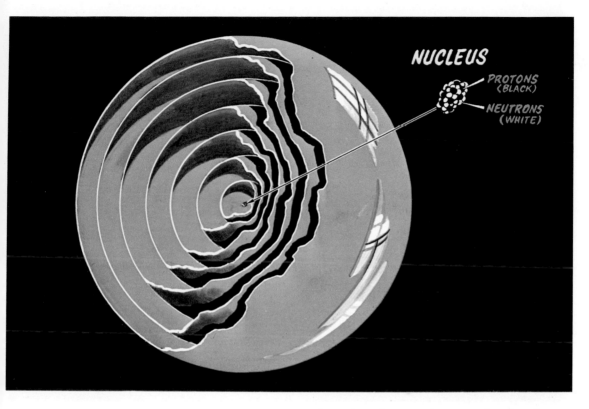

NUCLEUS

PROTONS
(BLACK)

NEUTRONS
(WHITE)

A section of a complex atom showing the nucleus containing protons and neutrons surrounded by the electron shells. Like a planetary system, an atom is mostly empty space.

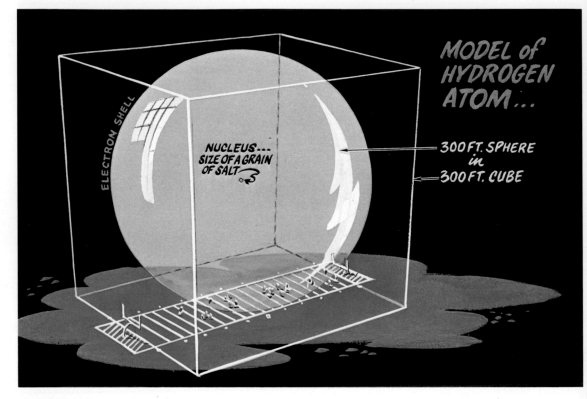

A model of the hydrogen atom. If the nucleus is made the size of a grain of salt, the electron shell would touch the sides of a 300-foot cube, or box.

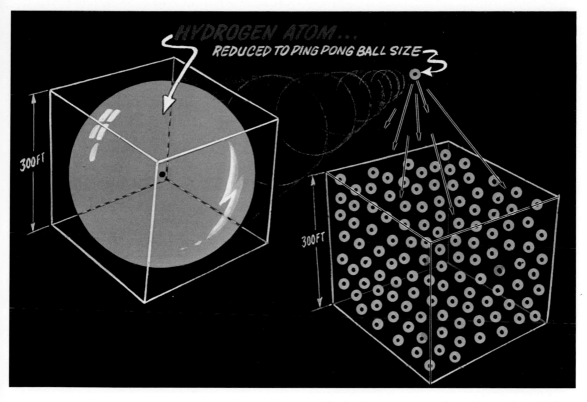

The hydrogen atom reduced from the 300-foot box to Ping-pong ball size.

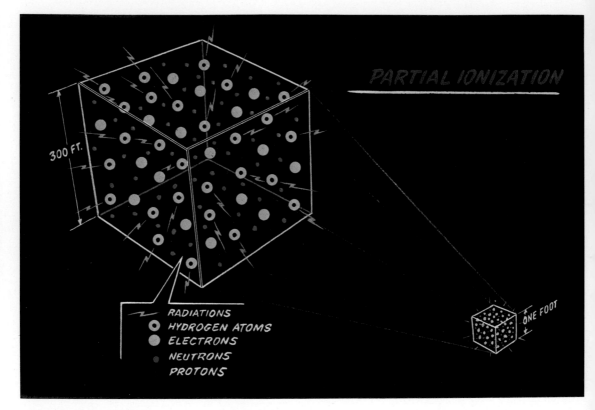

By subjecting the 300-foot box of Ping-pong-ball-sized hydrogen atoms to energetic radiations, the atoms are broken up (ionized) into electrons, neutrons, and protons, and the material in the 300-foot box can be squeezed into a box one foot on each side.

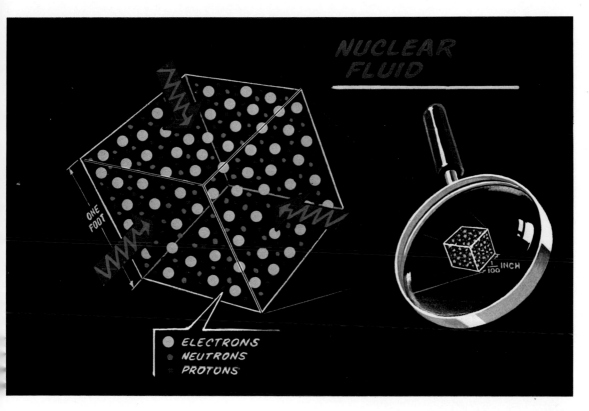

By continuing to subject the box to high-energy radiations, all of the hydrogen atoms become ionized to form a nuclear fluid which can fit into a box less than $\frac{1}{100}$ of an inch on each side. Thus, all of the hydrogen in the 300-foot box has now been squeezed into a box half the thickness of a fingernail on a side.

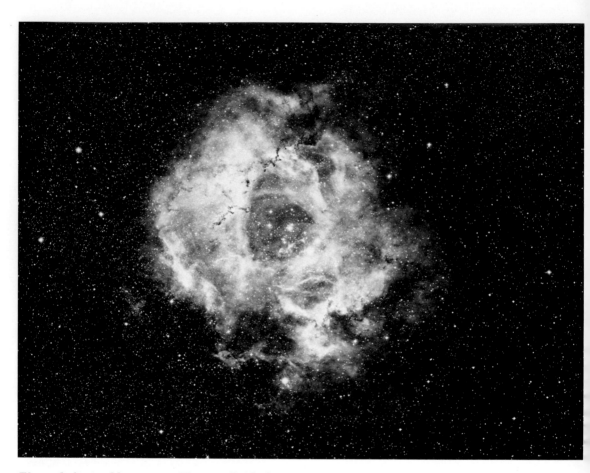

The nebula in Monoceros. The small black globules in the nebula represent the spawning grounds for stars. These globules are composed of clouds of dust and gas, some several light-years in diameter. Eventually radiation pressure and gravitational forces will compress them to create stars. (*Mount Wilson and Palomar Observatories*)

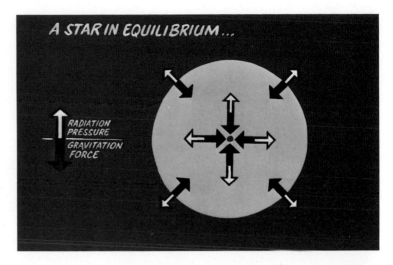

A STAR IN EQUILIBRIUM...

RADIATION PRESSURE

GRAVITATION FORCE

Once a star has arrived on the main sequence it is always in equilibrium. It remains in this state throughout most of its life. At any point in the star, from the inside to the outermost atmosphere, the gravitational force is precisely balanced by radiation pressure. This equality of forces insures the stability of the star.

RADIATION FROM SURFACE

CONVECTIVE ZONE

RADIATIVE ZONE

HYDROGEN BURNING SHELL

ISO-THERMAL HELIUM CORE

A section of a star on the main sequence showing a helium core surrounded by a hydrogen-burning shell and a hydrogen envelope. Radiation from the shell is transferred to the upper shell of the star by a radiative process, by convection into the outermost shell of the star, and by radiation from the surface. (*Adapted from* Essentials of Astronomy *by Motz and Duveen, courtesy of Columbia University Press.*)

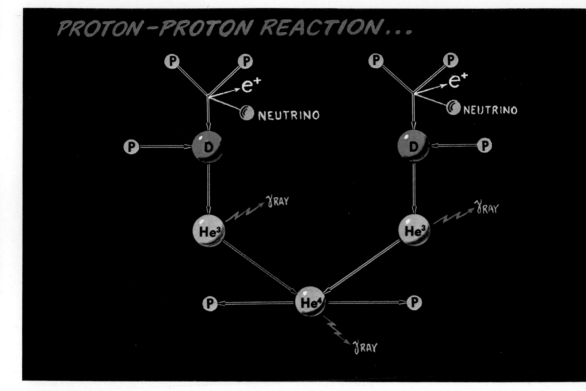

Proton-proton reaction. The fusing of two protons leads to the formation of deuterium. Deuterium, combined with a third proton, creates the light helium isotope. When two light helium atoms interact they fuse to form a normal helium atom, plus two protons which can enter into the fusion reactions once more. The difference in mass between what enters the reaction and what emerges is transformed into energy in the form of gamma rays and neutrinos.

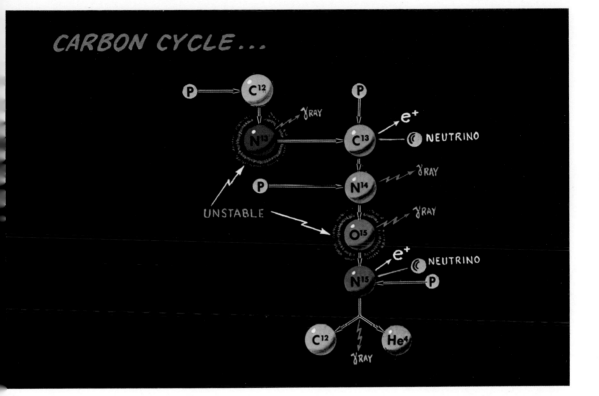

The carbon cycle. In this set of reactions four protons interact in six steps with various atoms to create more massive atoms. When nitrogen 15 is reached, the addition of another proton creates a condition which gives rise to a helium atom and a carbon atom. The carbon atom can now once more assume the role of catalyst for energy generation. The difference between the mass of the four protons and the helium atom appears once again as energy in the form of gamma rays and neutrinos.

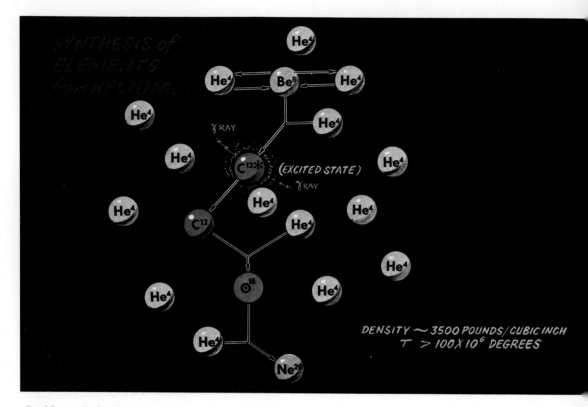

Buildup of the heavy elements. Two helium atoms can create a beryllium atom with a very short life span. If, before it disintegrates back to two helium atoms, it can capture a third helium atom, then a relatively stable carbon atom will be created. The carbon atom can capture still another helium atom to create oxygen which, in turn, can capture still another helium atom to create neon. In this fashion, the heavy elements are built up.

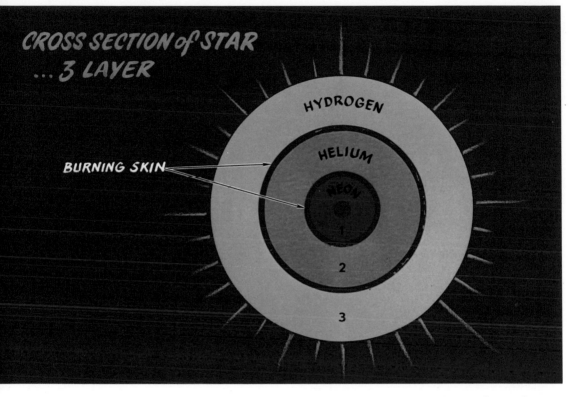

CROSS SECTION of STAR
... 3 LAYER

HYDROGEN

HELIUM

NEON

BURNING SKIN

1

2

3

Cross-section of a star at an advanced stage showing the neon core, the helium and hydrogen layers. The helium and hydrogen burning skins are indicated.

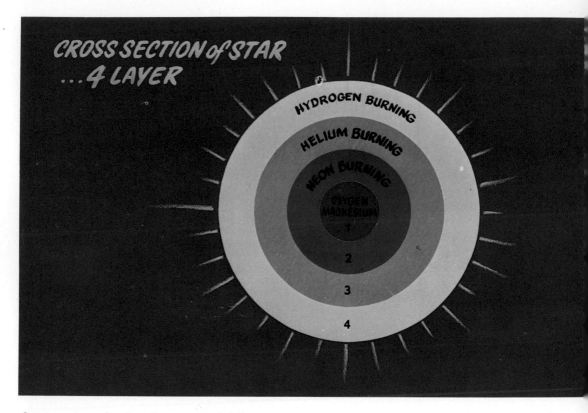

Cross-section of a four-layer star. As the star grows older, the neon is transformed to oxygen and magnesium. The burning skins, indicating the burning of hydrogen, helium, and neon that provide the source of energy for the star, are also shown.

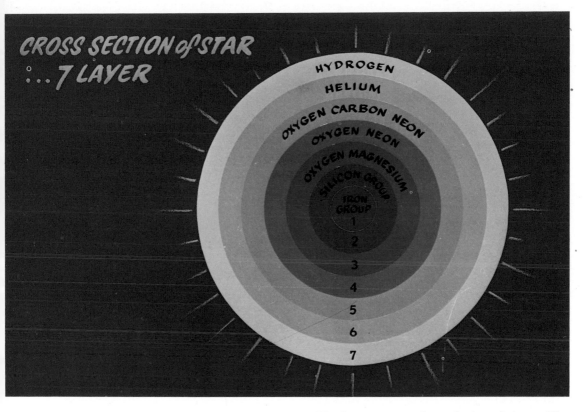

CROSS SECTION of STAR
... 7 LAYER

HYDROGEN
HELIUM
OXYGEN CARBON NEON
OXYGEN NEON
OXYGEN MAGNESIUM
SILICON GROUP
IRON GROUP
1
2
3
4
5
6
7

The last stages in the evolution of a star. The core now contains the iron group of elements surrounded by shells of silicon, oxygen, magnesium, carbon, neon, and helium. The outermost shell is still hydrogen.

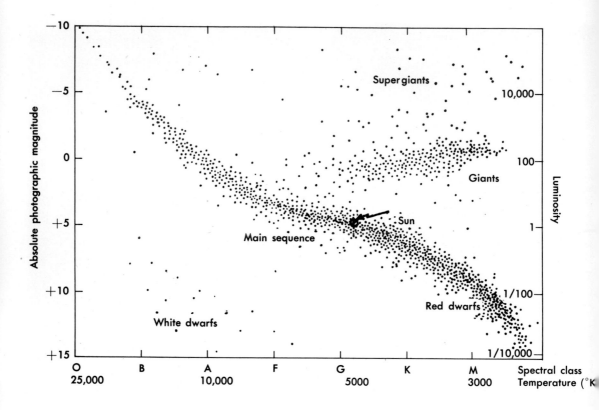

Most of the stars we see in the sky are normal stars. If their intrinsic brightness is plotted against their temperatures they would fall on a slant line that the astronomer calls the main sequence. The exceptional stars would fall outside this region. The place where a star initially joins the main sequence to begin its life span is determined by its mass. Stars of small mass begin at the bottom of the main sequence while stars of high mass begin their life span high on the main sequence.

If we encounter a cluster of stars in the sky and want to know how far away it is, all that is necessary is to find a cepheid variable and determine its period. Then we measure the star's brightness on a photographic plate. From the period we know how bright the star should be and from the photographic plate we know how bright the star is, and, with this knowledge, we derive the distance. Using the brighter cepheid variables we can move out to galaxies of stars that are millions of light-years away because these cepheid variables are, on the average, about 5000 times brighter than our sun. For this reason, galaxies containing these variables may have their distances determined out to somewhat more than 10 million light-years.

Before leaving the cepheids it should be noted that the calibration of the cepheid variables has undergone some modifications to bring them into step with reality. In 1952, Walter Baade proved conclusively that there were at least two types of cepheid variables. While the shorter-period cepheids—they are the most plentiful in our neighborhood—apparently do conform to the above distance criteria, it was shown that the longer-period cepheids were actually twice as distant as previously indicated. The net result of this discovery was that distances in the Milky Way remained the same but distances to other galaxies were expanded by a factor of two.

What happens when the galaxies are so distant that we can no longer detect the cepheids? Now the astronomer uses clever stratagems which involve the supergiant or brightest stars and the novae. The supergiant stars are on the average about 75,000 times more luminous than the sun and, if we can recognize these in the galaxies, we can again determine distances, by comparing how bright they should be with their photographic brightness. This technique has been used to probe the depths of space out to distances in excess of 20 million light-years.

Novae (which means new stars) are not new stars but old ones that suddenly flare up and become tens of thousands of times brighter than they were previously. They appear to reach a peak luminosity of about 2.5 times that of the supergiants. Thus, we can use the average nova to penetrate space to a depth of over 30 million light-years to determine distances to galaxies. The typical nova will flare up in a day and in some cases take years to return to its former, faint state. Novae, because of their presence, permit us to penetrate deeper into space to determine distances.

The next yardstick represents the globular clusters which comprise compact systems of 100,000 stars or more and which ring our Milky Way system in a spherical cloudlike pattern. These globular clusters can achieve a luminosity 6 or more times that of the novae, which permits us to probe space to a depth of over 80 million light-years. At this distance, we are reaching clusters of galaxies.

The next step involves the use of supernovae, which we will meet in great detail later. These stars flare up with a brightness that can rival the luminosity of a galaxy of scores of billions of stars, and they permit us to probe space to a depth of about 1 billion light-years. In this determination there is a problem; the intrinsic brightness range in the supernovae is 3 or 4 magnitudes, so that the distance determination can be off by a factor of 6. They cannot be used for precise distance determinations to the galaxies. All we derive is a "ball-park" figure. But even with this yardstick, we are not yet approaching the edge of observable space. How can we proceed? At the present time, the only recourse is to use the light of an entire galaxy. We can use the light of giant galaxies to push our horizons back to almost 10 billion light-years. However, when we get involved with galaxies, we can use a second method of determining distances to the most remote objects. For the reasons behind this method, we must retrace our steps and engage in some historical research.

We discovered, at the conclusion of Chapter I, that the Doppler shift could indicate to us the approach or recession of celestial objects. In the years between 1912 and 1925, V. M. Slipher obtained spectra of many galaxies and discovered that the spectral lines were shifted toward the red end of the spectrum. He discovered that there was some relationship between the red shift and the size and brightness of the galaxies. However, shortly thereafter, these red shifts were considered indicators of velocity and, if the Doppler shift was interpreted correctly, all the galaxies appeared to be retreating from us (if local motions were taken into account and eliminated) and the fainter objects (hence, more distant) appeared to have a larger red shift, indicating that they were retreating from us faster. By 1929, Edwin P. Hubble had determined new estimates of the distances to many galaxies whose velocities had been measured and found that the velocity of recession was directly related to their distance. The larger the red shift, which meant the faster they were moving away from us, the farther away they appeared to be. This was firmly established in the early 1930s, and to this day this velocity of recession has been used as a rather flexible yardstick to measure the distances to these remote objects.

We call it flexible because the scale of this yardstick has been changed from time to time as astronomers involved with this problem expanded the known universe. We are not certain that the scale in use is an absolute one. Currently, the Hubble constant which specifies the rate of recession with distance is given as from 13 to 30 miles per second for every million light-years of distance. This distance-measurement concept is an unusually important element in our story; later we use this method to determine distances to the mysterious quasars.

A significant point which involves our dependence on distance determination must be introduced here. It is most essential that distances to the most remote objects be determined reasonably correctly if we are to move forward in our understanding of the universe. We must, however, become reconciled to the fact that distances which are determined will contain errors—some of them of major magnitude. What must be remembered is that we have to live with this uncertainty, for we have no other choice. In time, some of these determinations will be modified as, indeed, the cepheid-variable scale was modified. The important point is that we must have some yardstick, and today we have one and use it—even with its uncertainty.

The time has now come to talk of the brightness of celestial objects. Some of our story is closely tied to brightness or (to be more precise) luminosity. However, when we talk about the objects we see from the earth, we are concerned, first, with how bright they appear—either to the naked eye or through a telescope.

Some stars are easily seen, even from an unfavorable location such as downtown New York. Others can be seen only by concentrated effort from the most ideal observing spot on earth. Many more can be seen with a small telescope, and we are certain that others exist which are not discernible through a small telescope.

The range of apparent brightness, from the sun to the faintest object detectable with the largest telescope, is tremendous. We can say:

> *The sun is 10 billion times as bright as Sirius, the brightest star in the night sky, which is about 1000 times brighter than the faintest star visible under ideal conditions, which is about 6 million times brighter than the faintest object photographed in the large telescope. Thus, the sun is about 60 billion billion times brighter than the faintest galaxy that can be detected.*

These brightnesses are related to one another in a way which has come down to us rather unchanged over a very long period of time.

Hipparchus (about 135 B.C.) chose the brightest stars, on the average, as stars of the first magnitude. The faintest star visible to the naked eye he called the sixth magnitude. This scale was used for more than 2000 years, or until an astronomer decided to study its significance. Then it was discovered that one of the average stars of the first magnitude was about 100 times brighter than a star of the sixth magnitude. But there are some stars brighter than the first magnitude, so we must go to zero magnitude and even to negative values. Sirius, for example, is 10 times brighter than a first-magnitude star. Thus, Sirius has a negative value of minus 1.6 magnitude.

While a range of 5 magnitudes is the equivalent of 100 times in brightness, one magnitude is not 20 times, for the magnitude scale is a geometric one. Thus, a first-magnitude star is 2.512 times brighter than one of magnitude two; a second-magnitude star is 2.512 times brighter than a star of magnitude three; and so on, until a star of the fifth magnitude is 2.512 times brighter than a star of magnitude six. If one multiplies $2.512 \times 2.512 \times 2.512 \times 2.512 \times 2.512$ the answer is 100. Thus, a sixth-magnitude star is 100 times fainter than one of the first magnitude.

The last element we will dwell on before beginning our discussion of the paradoxes deals with the energies of various systems in the universe. To understand what is going on in these systems it is necessary to have some knowledge of the units in which energy is expressed and in what magnitudes.

Physicists define a unit of energy—the erg—as the kinetic energy or the energy of motion of a mass of 2 grams (about 1/14 of an ounce) moving with a speed of 1 centimeter per second (about $\frac{3}{8}$ of 1 inch per second). How much energy is this? The late George Gamow indicated this is the energy exerted by a mosquito flying across a room. Thus, the erg represents a tiny amount of energy. When we speak of the energy output of the sun or stars, we are speaking about vast amounts of energy. We, on the earth, receiving about one two-billionth the energy emitted by the sun, find that this energy can create clouds, evaporate vast amounts of water, heat homes, and do many more things, for we receive about 15 million ergs per second on each square inch of the earth. If this is what the earth receives, it means that the amount of energy the sun emits must be colossal. It is staggering! When we compute the energy output of the sun, we find it is 3800 million million million million million ergs per second. If we were to write it out it would look like this: 3,800,000,000,000,000,000,000,000,000,000,000 ergs per second. To try and remember it, and to write it, would be time-consuming and we might forget a zero or two, so the scientist has developed another shorthand which he uses. Just as the astronomer uses a light-year to measure distances to stars that are millions of millions of miles distant, so the physicist uses a notation which makes life a bit easier for him. The physicist uses "exponents" to describe large numbers.

As an example, we can start with 10. This could be written as 10^1 which means it is 10 to the first power, or 10. If we write 10^2 it means that this is 10×10, or 100, that is, 1 with two zeros following it. In the same fashion 10^3 would mean $10 \times 10 \times 10$, or 1000, or a one with three zeros following it. We can, thus, quickly write any number containing a large number of zeros in this short-hand notation. Now we can say the energy emitted by the sun is 3.8×10^{33} ergs per second. This number was written out above. In this notation there can be multiplication and division and the chances of falling into errors are diminished by its use. We will find it used extensively in the body of the book.

3

Of Stars and Models

Analogies and models are often used to try to make concepts and ideas understandable to the layman. Because the problems of dealing with matter—when it is concentrated to the extreme densities of the centers of stars—are so alien to normal terrestrial experiences, it is almost impossible to discuss them with a layman without the use of a model. For this reason, a concerted effort will be made to formulate a model for the pictorialization of matter at extreme densities.

To begin, imagine a large, empty box in orbit around the earth. The box is a huge cube about 300 feet on a side (it could contain a thirty-story building in any position). As the box is in orbit around the earth, it is subject to gravity-free conditions and no forces are acting upon it. The "falling" of the box around the earth compensates for gravity. Thus, someone roaming around the box would be weightless and behave like an astronaut in orbit around the earth or moon. Let us further imagine that it is possible to push in the sides of the box to make it smaller and smaller as the occasion demands. The pushing on the sides of the box may be considered an irresistible force. Thus we start off with a box in space—under free-fall conditions, with the vacuum conditions of space, and sides that can be pushed together to reduce the volume.

I am going to fill the box with very special light bulbs. The normal light bulb has a small filament in the center and a glass envelope around it. Also, the inside of a normal light bulb is considered a fair vacuum. But, I would like to use a special spherical light bulb, which I would have made up for me by some skilled artisans. I would insist that the filament be a tiny kernel in the center of the bulb similar to that of the high-intensity lamps currently in use. I would further insist that the kernel be made up of tiny black and white spheres closely packed together. The black spheres I will designate as protons and the white spheres as neutrons. Instead of a single glass envelope around the kernel, or filament, I would like to put several concentric glass envelopes around the filament—seven to be precise. This will be a very special glass, for when the glass envelope is broken the glass would instantly disappear except for a small moving speck we call the electron (remember Dr. Goudsmit's explanation of the electron). Each light bulb I will consider an atom, with the kernel as the nucleus and the glass shells, or envelopes, as electron clouds which circulate around the nucleus. Thus, I have constructed a model of an atom. But now we must talk about sizes! What size shall I make the atom? What is the relation of the kernel to the glass envelope? Let's consider the hydrogen atom.

If I choose to make the nucleus the size of a grain of salt, then the diameter of the outside shell is about 300 feet—the length of a football field. From this, one readily sees that the atom is really a huge volume of empty space. In a 300-foot sphere (the height of a thirty-story building) I would find in the center a grain of salt and circling around it at a distance of 150 feet would be the electron shell. But, I am beginning with a visible grain of salt to represent the nucleus. In reality, the atom is so small that I would have to squeeze in 100 million of them shoulder to shoulder to cover the width of the fingernail on my little finger. Further, the nucleus in the center of the atom is so small that 10 trillion of them are necessary to cover that same fingernail. To appreciate the infinitesimal size of these, I would have to fill 27 billion spheres the size of the earth with these 300-foot hydrogen atoms to make them add up to one ounce! And, most of this weight would be in the nucleus, for it takes the mass of 1837 electrons to equal one proton.

Now that we have some idea as to the scale of the hydrogen atom, let's shrink it from the size of a football field to a ping-pong ball.

I am going to pack that earth-orbiting box with hydrogen atoms. Because they are individual atoms, they will behave as a gas and move around in the box with relatively high velocities. If we focus a beam of light on the atoms, many of them will absorb some of the radiation in the light beam and move more rapidly. If we concentrate and focus the light on the atoms, they will absorb still more energy and move even more rapidly. If ultraviolet radiation, or X rays, which contain much more energy, is used, the atoms will move still more rapidly. Finally, we might irradiate them with so much energy that their motions become fast enough for them to begin shedding glass envelopes. When this happens, and we examine the box, we will find some bare filaments and some tiny specks—the remains of the empty glass envelopes wandering around. The filaments represent the nuclei of the hydrogen atoms, or protons, and the specks represent the electrons which circled the nuclei. The hydrogen atoms under these conditions are ionized. As was previously indicated, a characteristic of ionization is that the proton has a positive charge and the electron possesses a negative charge. Because they are now charged particles, they are subject to magnetic fields—this phenomenon is used to make motors operate—and they are subject to certain physical laws in which like charges repel each other and unlike charges attract. But, let's go back to the box!

Because unlike charges attract, the electrons may combine with protons to form neutral hydrogen atoms and the hydrogen atoms may again be broken up by the agitation impressed upon them by radiative energy. Thus, there is, under certain conditions, a continual breaking up, or dissociation, and a recombining of the atomic components.

Now the distance between the nucleus and the electron cloud is roughly 100,000 times the diameter of the nucleus. If the hydrogen atom is the size of the ping-pong ball, its nucleus is so small that about 5000 of them will have to be put together to cover the period at the end of this sentence. This means that when atoms are ionized and the electron clouds no longer impose large boundaries on the atom, they can be packed into a tiny volume. But, it must be stressed that this is only true when the atom is ionized. If in some fashion we could strip all the electrons from the atom so that only nuclei remain, we could, if the nuclei remained stationary, pack the nuclei so that they could just touch (this is not possible, but for the sake of our discussion let us assume this could be done), then we could put roughly 1 million billion nuclei in the space formerly occupied by a single atom Thus, if ionization could be carried out to affect all atoms in a given volume, the volume could shrink to incredibly small dimensions, and it would become incredibly dense. But let's get back to our box with the light-bulb atoms.

For atoms other than hydrogen there are more than one electron shell. These atoms can also be affected by the absorption of energy. If we assume that we have a mix of atoms and these are ionized to shed their outer shells, the ionized atoms can be packed closer together to reduce the volume with an increase in density. As a consequence, the number of atoms with which we began originally can be packed into a fraction of the volume of the box. We have in effect concentrated the numbers, packed them closer together, and have increased the density.

Again, let's pour some energy into the box. This time, let's imagine the box is hit by a meteoroid. The energy in the meteoroid is considerably higher than the other form of energy used earlier. The result is that a second glass envelope peels off, and the atoms become still smaller. These electrons also move about the space the atoms occupy—and again, they may recombine with atoms and may again be stripped from them. Thus, there is a dissociation and recombination process going on in this phase. But, in losing another electron shell, the size of the atom is further reduced and, as before, the atoms can be packed into a smaller volume of the box. The number of atoms will remain the same, but the volume will have decreased and the density, as a consequence, will increase.

Once more, energy is transmitted to the box. Now let's imagine a large meteoroid hits the box. Another shell of electrons will be stripped from the atom and the electrons will begin flying about, joining their friends. The atom is now becoming highly ionized, for it is losing many electrons. At this stage it takes many electrons to neutralize the excessive positive charge on the nucleus. This is theoretically possible but now it is rare that the full electron complement can be returned to the atom. The electrons and nuclei move around very rapidly in a small region in the box, thereby giving it a still greater density.

Let's assume that a swarm of many meteoroids strikes the box. These meteoroids penetrate the box and literally shatter all the electron shells and strip them all from the atom, except those in the innermost electron shell. With this stripping the box contains a mélange of electrons and nuclei moving with incredibly high velocities, for the pressure has become so high that it is the only way in which they can resist being totally crushed.

The scientist calls this process of separating electrons from their nuclei *pressure ionization*. When this stage is reached, the electrons cannot be identified with any particular nucleus and the nuclei can come as close as the innermost electron shell. If we now examine our box, we find that we have a sea of electrons—we can call it an "electron gas," moving about in a much heavier and more sluggish gas of positive nuclei so that the material in the box has the general characteristic of a metal. By this, we mean that if the sun shines on one side of the box, the heat is transported to the other side by conduction, that is, the atoms pass it on, one by one.

The scientist calls this electron gas "degenerate," for the electrons cannot be brought to complete rest even though the gas may be cooled to the absolute zero. There is a principle in physics, called the "Pauli exclusion principle," which states that no two electrons, if they are spinning in the same direction, can occupy the same element of space. If they are spinning in opposite directions, then no more than two electrons can occupy the same element of space. For this reason, the electrons may be moving with speeds which indicate a temperature of millions of degrees but the temperature of the electrons as a whole is very low. If you stuck your hand in it you would not get burned.

If the volume of the gases in our space box is compressed under higher pressures (in a star this would correspond to gravitational contraction), the electrons could move with speeds approaching the speed of light. Under these conditions, the electrons in some mysterious fashion tunnel into the nuclei. This sets off a chain of events. The electron will combine with a proton in the nucleus and convert the proton into a neutron, and that most curious particle—the neutrino—will be emitted.

The neutrino, as we have discovered in Chapter I, is one of the ghosts, or phantoms, of nuclear energy. This elusive particle has neither an electric charge nor a rest mass. Its sole function appeared to be to make the nuclear reaction equations balance out. The neutrino is so elusive that it takes a special set of experiments to detect it because of its extremely low level of interaction. One neutrino in millions can be detected by an interaction and, as we have discovered, the distance they can move without interacting is on the order of many light-years. But the neutrino becomes vitally important to scientists trying to understand what takes place in the cores of stars because neutrinos are the particles which remove vast stores of energy and under extreme conditions will heat up the outer shell of a star to initiate a cataclysmic explosion.

Getting back to the electron interaction with the proton to convert them to neutrons: this process will continue until the nuclei become unstable. By this we mean that the nuclei will contain too many neutrons compared to protons, and when a critical balance is reached the nuclei will sunder into individual particles. Now the material in the box will consist of a mixture of three degenerate gases: a proton gas, a neutron gas, and the electron gas we started with. There is still considerable room between the particles, but they are much closer together than when we started. Actually, the original volume of the giant, orbiting box has been compressed to about 1 cubic foot and the material in the box is 1 million times as dense as iron. But, still more pressure can be applied to the box. In the core of stars this pressure is created by gravitational contraction but in our box we are simply using inexorable pressure on the walls to shrink the box. The box continues to grow smaller until the density is 10 trillion times that of iron. Now, that one-foot box has been compressed until it is a tiny cube with its side the thickness of your fingernail. All of the mass in that giant box, 300 feet on a side, has shrunk to a tiny cube less than 1/100 of 1 inch on a side. The material is now a nuclear fluid. The many light bulbs with which we filled the box have been compressed to the fundamental particles of which the universe is composed. The bulbs have been compressed almost to the point where they disappear. It is by the complete disintegration of the atoms that matter can be compressed almost beyond limit to densities almost beyond comprehension.

4

The Evolution of the Stars

To understand what happens to stars as they pursue their lives and grow old, one must also understand how stars are born. While in the past this presented a mystery of monumental proportions, today's astronomers can detail with considerable assurance the steps that reveal how these celestial objects become the brilliant stars of our night sky.

Not too long ago astronomers were of the opinion that stars took many millions of years to evolve from the dust and gas of interstellar space. However, quite recently, startling and dramatic photographs have been taken of a region contained within the Great Nebula in Orion, where, in the course of a few years, a small cluster of stars apparently came into being. On a photograph taken in 1947 a group of three stars was seen. By 1954, several stars in the group appeared elongated, and, by 1959, the stars had separated into individual ones—making it the first time in the history of man that stars were actually seen as they were being born. This unprecedented event indicated to astronomers that stars are born in relatively short periods of time and the previously held "wild" speculations, that they normally arise in groups, or clusters, turns out to be correct. What is the mechanism by which they arise? Why, in the many years during which astronomers have been photographing and studying the skies, is this the first time that stars have been seen to materialize? The birth of stars cannot be a rare event, for all over the sky can be found the spawning grounds for these bodies.

When photographs of the nebular regions of our Milky Way

are scrutinized, there can be seen small, black, irregular patches, or globules, which represent massive collections of dust and gas. They are seen as black, for they have no light of their own and they lie between us and the bright background stars, thus obscuring the light of the stars behind them. As they are clouds of dust and gas, they contain dust particles that are most efficient in absorbing light coming from the background stars. These globules are of tremendous size, some being several light-years across. Even though the material in the globules is thinly strewn, the total volume is so enormous that enough material is contained within them to form a small cluster of stars, each as massive as our sun.

To visualize how stars arise from globules, we must realize that all stars in the sky emit radiation and that this radiation exerts pressure. Delicate instruments have been developed and used on earth and they react to the radiation pressure of the sun's light, even though this radiation has been filtered by the earth's atmosphere. In the case of a black globule, we find that radiation pressure from the surrounding stars tends to compress it into a more compact mass. In the globule, winds which blow the dust particles and gas in every conceivable direction are present so that the material in the globule exists in a violently turbulent state. Thus, the globule may be considered a turbulent mass of dust and gas being subjected to radiation pressure from all sides. The result is that the dust and gas will be compressed into a smaller and smaller volume. Depending on the source and intensity of the radiation surrounding the globule, the contraction will take place in a finite time and, because of the mass of dust and gas, a gravitational force will be generated in the center of the globule. Gravitational forces will begin compressing the globule by forcing the material in the globule to fall inward, toward the center. As a consequence of this fall to the center, kinetic energy will be given up by the particles, and the cloud of dust and gas begins getting warm.

For perhaps hundreds of years, this material will fall. At first, it will fall slowly and deliberately because the force of gravity pulling it in toward the center is quite weak. However, with the passage of time, the globule will grow smaller, the gravitational field will become more intense, and the fall will accelerate. But, as we have seen, the globule is big, on the order of one or more light-years in diameter. This means that the distance from the outer edges of the cloud to the center may be in excess of 6 trillion miles. If a particle were moving inward at the rate of only 1 mile per second, it would take about 200,000 years for it to reach the center. Observations indicate that the speeds of the dust particles and gas will be considerably greater than this, therefore the time scale for the gravitational compression will be much shorter.

The inward fall of the particles gives rise to many collisions between the particles, transforming kinetic energy into thermal energy, and resulting in an increase in the temperatures of the globule. We now have a protostar. It is shining because the energy of motion has been converted into heat to warm the gas and dust. At this point, the protostar is barely visible, for it emits most of its energy in the far infrared. The important point is that a star has been conceived but not yet born. Astronomers do not know how long it takes for the protostar to reach the stage where it begins to glow as a dull-red-color ball and becomes visible. Various estimates of this time scale range from thousands of years to several million. However, in light of the appearance of the stars in the Great Nebula in Orion, this writer would speculate that the shorter time scale may be the one closer to reality.

At this point we must pause for a moment to consider some aspects of stellar obstetrics and its bearing on the future life history of the star. Stars are created with a variety of masses, or amounts of material in them, along with a variety of the elements. Both of these factors influence the future behavior and life history of the star. To understand what happens, let's imagine ourselves going out of doors to look into the night sky.

From a mountaintop, far from the disturbing lights of a city, you may see as many as 3000 stars in the sky. Under the most ideal conditions, and with the keenest of eyes, you might see half as many again. Some of these stars are as close as a few light-years, others are at distances up to 1000 light-years. Now, suppose we could take all these stars and plot them on a diagram. The two characteristics to which we will pay particular attention are the temperature of the star surface and the intrinsic brightness—its luminosity. If we plotted the 3000 stars, we would find that the brightest stars were also the hottest. The faintest stars would be the coolest. We would discover that almost all the stars would fall along a slant line which runs from the upper left to the lower right of the plot. These represent the normal stars in the sky and they fall along this slant line, or path, the astronomer calls the "main sequence." In reality, it is called the Hertzsprung-Russell diagram, named for the two perceptive astronomers who first evolved this significant relationship. In this relationship, mass plays a significant role. If the mass of a star is high, it will, on birth, enter the main sequence at the upper part of the slant line. If the mass is small, it will enter the main sequence near the bottom.

The significance of the main sequence is that most of the stars we see in the sky are ordinary stars without special characteristics and, thus, we should expect to find them fitting a certain pattern and this pattern is, indeed, the slant line. The reason why they are on the slant line is that although it may take only a few hundred thousand years to enter the main sequence and, while from the time it leaves the main sequence to the end of the life cycle may take as long as a few hundred million years, most of the stars will reside on this main sequence for literally billions of years. Thus, the birth and death of a star is like the wink of an eyelid on the cosmic time scale. In the case of the sun, which is an ordinary type of star, it has been on this line for about 5 or 6 billion years and may continue for another 5 or 6 billion years; for the lifetime of a star, with the mass and composition of the sun, is 10 to 12 billion years. Stars having considerably less mass than that of the sun will reside on the main sequence for perhaps 50 billion years. Those stars with 30 or so times the mass of the sun will reside on the main sequence for a million years or less.

Returning to the birth of the star, we find that the compression continues to increase and with this increase comes a rise in temperature. The temperature climbs higher and higher and the tremendous ball of gas and dust begins to glow and can be seen as a dull, reddish disk against the black background of the sky. A significant proportion of its energy is still in the infrared. But—this still does not constitute a star. As the material in the protostar becomes more and more compact, the material will fall faster and faster toward the center and, as a consequence, the core of the protostar grows hotter and hotter. Finally, the temperature in the center reaches 10 million degrees and this temperature triggers the thermonuclear reactions that constitute the energy-generating mechanism of all the stars in the universe. With these thermonuclear reactions, a full-fledged star has come into being.

When the dust and gas accreted to form the protostar, the sample of material which went into this was representative of the material in a particular part of the universe. A sampling of the content of interstellar space would disclose that, by weight, hydrogen represents almost 89 per cent of the materials there. Helium is a distant second, with 10 per cent, the elements such as oxygen, nitrogen, carbon, neon, etc., represent 1 per cent, and all the metals put together does not exceed $\frac{1}{4}$ per cent. Thus, the star is made from those elements which are most prevalent in the universe. Because hydrogen is the most plentiful element, it is certain that any thermonuclear reactions which take place must involve hydrogen. In some corners of space the distribution of the elements may show a higher percentage of the heavier elements, but these are local anomalies since these represent regions where, in the past, stars have exploded to scatter and diffuse the heavier elements in their immediate neighborhood. For the time being, we will ignore this special concentration of the heavier elements and concentrate on stars which are formed primarily of hydrogen.

When the temperature in the center of the protostar reaches 10 million degrees, a complex but well-understood set of fusion reactions takes place which builds up helium out of the hydrogen nuclei, or protons. This reaction will creat a helium atom by combining four protons. In this process, a proton will combine with another proton to form an atom of heavy hydrogen, or deuterium. When this is bombarded by a third proton, another reaction takes place to create the light isotope of helium, containing two protons and one neutron. In the hurly-burly realm in the core of the star, the swiftly moving light helium atoms collide with one another. As a consequence of this reaction, normal helium, containing two protons and two neutrons, will be created and the two remaining protons loosed in the hot mélange of the core to enter once again into reactions to create helium out of hydrogen. In this process, about 0.7 per cent of the mass will be converted to energy. This represents the important thermonuclear reactions which occur in the stellar cores at temperatures of about 10 million degrees.

Some astronomers speculate that other reactions can take place at somewhat lower temperatures involving lithium, beryllium, and boron. However, they also indicate that if these reactions occur their contribution to the total energy generated is minor.

As the temperature in the core rises, another significant reaction occurs which involves carbon as a catalyst. Beginning with hydrogen and carbon 12, the reaction entails the formation of nitrogen 13 which spontaneously disintegrates to carbon 13—an isotope of carbon heavier than the one we started with. The carbon 13 absorbs another proton, which will be transmuted to nitrogen 14. Nitrogen 14 absorbs another proton, to become oxygen 15. This element also is unstable and spontaneously disintegrates into nitrogen 15. Finally, nitrogen 15, with the addition of a fourth proton, breaks down into a carbon 12 and helium atom. Thus, a by-product of these thermonuclear reactions is carbon 12, which can initiate the same type of reaction all over again. Four protons were combined to form one atom of helium and the difference in weight between the four protons and the helium atom—about 0.7 per cent—appears as energy emitted by the star. In the case of the sun, in each second of time, 564 million tons of hydrogen are transmuted to 560 million tons of helium and the difference—4 million tons of matter—is converted and radiated away as energy.

This points up the significant fact that the energy-generation mechanism in a star is temperature-dependent. The core temperature of the star determines the process. Astronomers assume that at a temperature of 13 million degrees the carbon cycle is relatively unimportant. Thus, even at this temperature, the proton-proton reaction is dominant. Increase the temperature to 16 million degrees and both cycles probably contribute equally to the energy output. However, when the core temperature rises above 20 million degrees, the carbon cycle becomes dominant.

Once nuclear reactions are under way to provide energy for the star, the contraction which initiated the energy cycle is no longer necessary and thus stops. Now the self-sustaining reaction can continue for a period of from less than 1 million years to more than 100 billion years, depending on the initial mass of the star. It is at this point that the star has reached the main sequence and that its long life can begin, and continue with almost imperceptible changes. The "eternal stars" characterize this phase of a star's life. Nothing occurs which could single out this star to make it an object of careful scrutiny. It is simply a self-respecting member of the stellar community, lost in the anonymity of numbers. But, the processes going on in the core of the star contain the seeds of its own destruction.

When wood or coal is burned in a fireplace, the by-products are heat, smoke, and ashes. In the core of the star, hydrogen corresponds to the coal and helium to the ashes. As in a coal furnace, if the ashes are not removed, they clog the furnace and the fire can go out. In the case of the star, to avoid this condition, the core remains dormant and the shells surrounding the helium core now partake of the thermonuclear reactions that provide the energy of the star. However, in time, these shells also have their hydrogen consumed and the core grows larger and larger. Finally, a stage is reached where no hydrogen remains in the core of the star. Normal reactions converting hydrogen into helium cease, and the star leaves the main sequence to begin a relatively short (but exciting) career marked by extremely violent reactions.

When hydrogen becomes scarce and ceases to react, the source of energy vanishes. But, we have seen that a star is an exquisitely and delicately balanced mechanism in which the radiation pressure, like a wind blowing outward from the center of the star, is neatly balanced by the gravitational pull of the star. Thus, when energy ceases being generated, the radiation pressure will drop precipitously and gravitation will take over to collapse the star. Once more there occurs an infall of matter to the center of the star in much the same manner as the initial material collapsed to create the protostar.

The energy generated by gravitational collapse is enormously greater than the fusion reaction to create energy, and so, suddenly, the star finds itself being subjected to extraordinarily high internal pressures. The result is that the upper layers are heated and the star expands and grows in size until the outer layers grow sufficiently tenuous to permit more radiation to leak out. In the case of a star like the sun, it is believed that the star will become large enough to fill the orbit of Mercury. When the star begins to expand it moves off the main sequence and from what we have discovered this means that the star's days are numbered. From this moment on, there is only one place the star can go and that is downhill.

When the star collapses, the gravitational effect generates an enormous volume of energy to swell the star. While one might suspect that this will cause a drop in the core temperature, this is not the case. The temperature in the core of the star, contrary to what one would expect, rises sharply. In a relatively thin layer around the core, normal fusion is still taking place. The normal fusion generates more helium, which further enlarges the core. When the core contains about one half the mass of the star, it has been expanded to its maximum and its color has gone from white to yellow to red, for the surface temperature of the star is low.

The star now enters a new phase. The temperature in the core has risen until it is in excess of 200 million degrees. This represents the ignition temperature of helium. The helium reacts to create carbon. This is also an energy-producing process, for here we find three helium nuclei fusing into carbon with the end product weighing less than the initial ones. Once more, the radiation pressure which played such a major role when the star was on the main sequence assumes the role of countering gravity and the core of the star is prevented from collapsing. The star now begins its return to normal size and, as it does so, it concentrates the energy available and its surface temperature rises so that the reddish hue becomes white.

At this point, for some mysterious reason the star becomes somewhat unstable. Astronomers believe that the variable stars, stars that periodically vary in brightness, arise at this stage of a star's evolution, for, apparently, the shrinking processes do not develop smoothly and rhythmic oscillations appear to be triggered. At this stage, the star may pass through the nova phase, during which it can eject, violently, considerable quantities of its own substance into interstellar space. This substance would comprise an expanding shell, or envelope, so that the result would be a significant decrease in the mass of the star. Some novae recur, meaning that the one outburst is sufficient to achieve stability and that more outbursts, ejecting more matter, take place. But, in time, stability is achieved, oscillations disappear, and the star begins its long descent to the stellar graveyard. The star, even at this stage, is still capable of acting up and making a spectacular and dramatic display of itself. It may become a supernova—but, more of that later. The reason for this spectacular display deals with the amount of matter the star possesses at this point in its life cycle.

When we were discussing the nuclear processes going on in the core of the star, we discovered that the principal product of the nuclear reactions is helium—the ashes. We were speaking about temperatures of 200 million degrees. As more and more of the hydrogen is transmuted, the helium core grows larger. The hydrogen diminishes in mass, thus the production of energy from this source also diminishes. But—at a temperature of 200 million degrees, there is still another method whereby helium is used to create heavier elements to release more energy. In this other process, two helium atoms combine to create a beryllium atom which normally should disintegrate back into helium atoms. However, the temperature is so high and the speeds of the helium atoms so great that the beryllium cannot split up before three helium atoms combine to form an atom of carbon.

The process does not stop here, for now helium atoms can bombard carbon to form oxygen, bombard oxygen to form neon, and bombard neon to form magnesium. As this occurs, the temperature of the core remains the same yet it is too low for the formation of still heavier elements. The core begins contracting again until a temperature on the order of a billion degrees is attained and the heavier elements can be synthesized. If, because of further core contraction, the temperature goes beyond 3 billion degrees, the heavy nuclei interact with each other until iron is created and, at this point, the process stops. If helium atoms begin to bombard iron nuclei, instead of creating heavier elements, the iron nuclei will fission.

At this stage in its life cycle, the core of the star consists of iron surrounded by layers of nuclei of the lighter elements until the helium layer is reached and this is still surrounded by a last thin shell of hydrogen, serving as a fuel to create some energy. There comes a time when even that bit of hydrogen is consumed and this mode of energy-generation ceases. The other modes of generation also cease and the star is left without any means of supplementing its energy supply—which means that the star must die.

With the exhaustion of the supply of nuclear energy, the star can only contract and use gravitational energy to keep it shining. The star will continue to contract and glow brightly, using gravitational energy. As this energy is consumed, the star will grow fainter and change from white to yellow to red and, finally, it will stop glowing and will begin its ceaseless wanderings through infinite space as a small, dark, lifeless object. But, on its way to extinction, the normal star passes through the white dwarf stage. Let's explore the life cycle of the white dwarfs.

5

White Dwarfs

The story of the white dwarfs represents one of the most fascinating in the history of astronomy. It marks the first time that celestial objects were discovered to possess properties completely alien to those of our terrestrial experience. And, in all likelihood, the resolution of the mystery of the white dwarfs sparked the inquiries into the bizarre nature of matter that may be found tucked away in odd corners of the universe.

The history began in the early nineteenth century when Friedrich Wilhelm Bessel, in tracing the motion of the magnificently brilliant star Sirius, discovered that its path was really not a straight line but possessed a wavelike character. The straight-line motion, called "proper motion," was not uniform and the bright star appeared to shift its position ever so slightly from side to side. By 1844, some ten years after he began his observations of Sirius, Bessel had concluded that Sirius was accompanied by a second star which could not be seen but made its gravitational effects manifest by the predictable wavy motion of Sirius. Even more intriguing was the realization that if a dark companion existed, the period of revolution of the two stars around their common center of gravity was on the order of fifty years.

We move forward in time to 1862 and shift the scene from Germany to Cambridge, Massachusetts. Alvan Clark had become the premier telescope-maker in the United States and had been

commissioned by the University of Mississippi to construct a telescope with an 18½-inch-diameter objective which was to make it the
largest telescope in the world. When he completed the task of
"figuring" the telescope lens it then had to be tested to insure the
accuracy of the figure. To do this, it was mounted in a make-
shift tube and turned toward Sirius, because it was the brightest
star and the best test for any defects. With the telescope tube fixed,
Alvan Clark saw a faint "ghost" come into the telescope at the
eastern edge of the field, in the glow of Sirius. As the sky continued to turn, Sirius burst into view. He was disturbed, for the ghost
could have represented a defect in the lens—a defect which would
have to be corrected before the lens could be accepted. However,
this faint star turned out to be the companion of Sirius, which
Bessel had predicted. To complete this story it should be added that
because of the Civil War, this telescope never traveled to Mississippi
but was installed at the Dearborn Observatory, near Chicago, and the
lens is still in use today—but in a new mounting.

Sirius thus became an object of vital interest and study, for the
physical characteristics of the binary system represented a dramatic
challenge to the astronomer. From the details of the motion of Sirius,
the distance from the earth, and the amplitude of the deviation from
straight-line motion, astronomers were able to determine the peculi-
arities of the two stars in the system called Sirius A and Sirius B.
The sum of the masses of the two stars turned out to be 3.4 times
the sun's mass. The distance between the two stars was found to be
about 20 times the earth-sun distance (which roughly corresponds to
the distance between the sun and Uranus). From the orbit characteristics, Sirius A turned out to have 2.5 times the sun's mass and the
companion, Sirius B, had 95 per cent of the sun's mass. When the
luminosity of the two stars is measured, it is found that Sirius A is
about 10,000 times brighter than Sirius B. From the absolute magnitude of Sirius A we know it to be about 35.5 times more luminous

than the sun. A bit of arithmetic discloses that the sun is 300 times more luminous than Sirius B.

The luminosity of a star depends on the temperature of the star's surface and its size; that is, its diameter. The proximity of the companion to Sirius A posed an extraordinarily difficult problem in trying to determine the spectrum of the star so that its temperature could be derived. In 1915, using all the technical competence and skills available at the Mount Wilson Observatory, the world's largest at that time, the spectrum was successfully photographed. With this came the dramatic disclosure that the temperature of the companion was about 8000 degrees compared to about 5700 degrees for the sun. The companion was in reality hotter and this meant more luminous per unit area than the sun. Actually, a simple computation indicated that each square foot of the star was radiating about 4 times as much energy as the sun. Now, if the sun's surface is 300 × 4 times as large as the companion, the companion must have a diameter of about 25,000 miles. But the star had 95 per cent of the sun's mass. This meant that a tremendous amount of material had to be packed into an exceedingly small volume which, in turn, meant that the star must be very dense. When the arithmetic is performed, the result indicates that the density of the companion is on the order of 100,000 times that of water and a cubic inch would, on earth, weigh several tons. A pint of this material would weigh 50 tons.

This, then, is the story of the discovery of the first white dwarf, and now we pose the question: how is it possible to have matter so compressed that a cubic inch would weigh several tons? The answer is the story of the dense stars of which the white dwarfs form the first class.

We have discovered previously that the elements may be compared with electric light bulbs with the filament representing the core, or nucleus, of the atom and the glass envelopes representing the electron clouds around the nucleus. We have also discovered that short-wave radiations will strip the electrons away from the atoms and the mixture of materials inside the star will be composed of atoms, with electron shells missing or, more precisely, ions, neutrons, and electrons.

But when matter is subjected to high pressure and compressed to high densities, as in the white dwarfs, another type of pressure develops which is termed "degeneracy pressure." This results from the sheer squeezing together of the matter in the center of the star. It is the squeezing that determines the degeneracy pressure and not the high temperatures. As a consequence of this squeezing action, the atoms are so closely packed together that their electron shells begin to interpenetrate.

With the passage of time, the white dwarf continues to contract gravitationally with continued interpenetration of the electron shells until the nuclei are as close together as the radius of the smallest electron shell will permit. These lowest electron shells represent impenetrable barriers to prevent further squeezing. When this happens, the electrons can no longer be identified with any particular nucleus and they are free to move about at will. The process of separating electrons from their nuclei, as we discovered in Chapter III, is the pressure ionization. When pressure ionization is complete, a cloud of electrons moves about in a matrix of the heavier nuclei so that the material of the white dwarf assumes certain physical properties of a metal. Radiation from this material is transported by electrons to the surface by conduction in much the same fashion as heat is transported in a poker made of iron that is being heated at one end.

But the electron gas exhibits peculiar properties. As electrons continue to be squeezed together, they acquire higher and higher velocities, for, as we have seen, there is a physical law dictating that no two electrons in any small volume can have precisely the same energy. Thus, they move at tremendous velocities to obey this dictate and avoid occupying the same element of volume. How large the smallest possible volume is, depends on the total range of electron velocities. However, on the average, the slower the electron velocity, the larger the minimum volume it can occupy; thus, the fastest moving electrons occupy the least volume. While there are some electrons that travel at speeds which indicate internal temperatures of millions of degrees, the temperature of the entire electron group is low.

In the ordinary white dwarf is found a gaseous configuration consisting of a matrix of closely packed heavy nuclei through which moves a degenerate electron gas. Toward the surface of the star the degeneracy drops off, and at the surface itself the atoms are not completely ionized and some of the material is in the familiar gaseous state.

Knowing these physical characteristics of the white dwarfs, we can construct a representative model. To begin, white dwarfs have an atmosphere. An analysis of the white dwarf spectrum discloses that while the atmosphere may be only a couple of hundred yards in thickness, the astronomer can detect various familiar elements in it. Two types of white dwarfs are recognized—a cool and a hot species. In the hotter white dwarf the atmosphere does possess some hydrogen, though it is probably less than 1/20 of 1 per cent of the atmosphere. However, from the lines in the spectrum of these stars we know that hydrogen, helium, calcium, iron, carbon, and even titanium oxide molecules have been found there. In the cooler white dwarf the atmosphere is almost all helium, with perhaps less than one atom in a million being hydrogen. Actually, we really have minimal knowledge of the composition of a cool white dwarf. There is a wide range in the absorption lines which are related to the temperatures of the surface and differences in surface gravity. The surface temperatures of white dwarfs vary from a cool 5000 degrees to a rather hot 50,000 degrees. The lines seen in the spectra of the stars are directly related to these temperatures.

Beneath the atmosphere lies a region of nondegenerate materials in which there are relatively few free electrons. This layer represents about 1 per cent of the radius with a thickness of about 100 miles. While this layer may change with time, the diameter of the white dwarf remains constant at about 25,000 miles in diameter. As a rule, the diameter of the white dwarfs does not decrease in size once they have reached this stage. They behave very much like a cannon ball that has been heated to a high temperature; the cannon ball can change in temperature and emitted energy, but its size remains constant. What determines the final diameter? It is the mass. Theory indicates that the larger the mass of a white dwarf the smaller is its radius, with the minimum radius being about 6000 miles. Theoretically, if the mass exceeded 1.2 times the mass of the sun, and with the proper composition, the radius could become infinitesimally small. (But this is another story—that of the black holes—to be detailed in a subsequent chapter.) It is the pressure of the degenerate electron gas which stabilizes the stars against any further contraction and, even though the temperature may go from millions of degrees in the core of the star to zero, the diameter will not change. The star will go out as a dark body with the same diameter with which it began its phase as a white dwarf.

When we get below this upper layer, we go into a degenerate interior which is practically isothermal; that is, it maintains a constant temperature down to the very center of the star. This temperature is probably on the order of millions of degrees—a likely number would be 6 million degrees.

Now that we have some idea as to the physical make-up of a white dwarf, what keeps it shining? One thing is certain: thermonuclear reactions are ruled out. There is no hydrogen inside a white dwarf that can support this mode of energy generation. The only source of energy in it is thermal energy. The nuclei of the atoms are milling around wildly, for they are scattered throughout the degenerate electron gas. With time, the nuclei slow down and this is equivalent to a cooling process. The electron gas, which is unlike any ordinary gas we know on earth, is highly efficient as a conductor and the electrons conduct the thermal energy to the surface where the atmosphere radiates this energy into space. As a consequence, some astronomers indicate that the hotter white dwarf gradually cools off —very much like the hot poker removed from a fire. When the white dwarf has just entered this phase, cooling is rapid, but as the internal temperature of the star drops, so does the cooling rate. It is estimated that the luminosity of a white dwarf will drop to 1 per cent of the luminosity of the sun in the first few 100 million years. Finally, the white dwarf will disappear as a black dwarf—but this may take many trillions of years and many scientists conclude it is extremely doubtful that the universe is old enough to have given rise to black dwarfs.

Other astronomers indicate that in the initial phases, when the white dwarf is still hot, the cooling rate is rather slow. However, once the star has decreased in surface temperature to that of the sun, the cooling rate accelerates with the rush to extinction being a precipitous one. And in this process when the white dwarf interiors cool sufficiently they will freeze into a solid mass.

In either case, with the age of the universe being in excess of 10 billion years, there should be many more red dwarfs than white dwarfs. Astronomers know this and have instituted searches for red dwarfs. To date, these searches have been unsuccessful.

The masses of white dwarfs are not too well determined. They are only determined definitively when the white dwarf is a member of a binary system, as in the case of Sirius. But there are quite a few that are members of binary systems. The results of these few indicate that the masses of the white dwarfs are, in the three best observed cases, less than the mass of the sun with an error in this determination of less than 10 per cent. Those whose masses have been determined, appear to have masses on the order of half that of the sun. Theoretical considerations indicate that the limiting mass for completely degenerate nonrotating stars is approximately 1.2 times the mass of the sun. However, if the star is rotating, and in all probability it is, then masses up to several times the mass of the sun appear possible.*

The surface gravities of the white dwarfs are on the order of 60 to 70 times that of the sun. I weigh 150 pounds on earth. On the sun I would weigh about 2 tons. However, on a white dwarf I would weigh between 120 and 140 tons. Considering that the diameters of the white dwarfs are fairly constant and with the masses rather consistent, it means that all white dwarfs have approximately the same surface gravity.

There are many white dwarfs in the sky. At one time, they were considered a rarity but careful scrutiny of the Mount Palomar-Schmidt plates indicates that there are in excess of 1500 in the sky. Astronomers believe that it is a fair assumption that white dwarfs are being born at a constant rate; at least, in the past 5 billion years. It is possible that white dwarfs may constitute the most numerous class in the sky. The space density of white dwarfs can be derived and it appears that there should be about 100 of these stars within a sphere about 30 light-years in radius.

We began this section with the tacit assumption that all stars pass through their evolutionary sequence and finally evolve slowly and inexorably to the white-dwarf stage—and then to extinction. But, many unanswered questions come to mind when this facet of the evolutionary process is carefully examined. We ask: which are the parent stars? Do all stars become white dwarfs at the end of their evolutionary paths? If the answer is not in the affirmative, then how

---

* We are going to run into this curious situation in which the astronomers find that problems have solutions for both nonrotating and rotating objects.

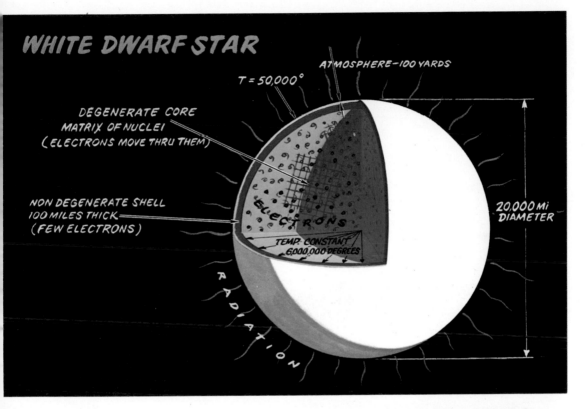

Section of a white dwarf star showing the very thin atmosphere, the nondegenerate shell, and the degenerate core which is crystallized with electrons moving through the matrix of nuclei. The temperature is constant through most of the star.

Crab Nebula in Taurus (photographed in red light). Arrow points to the pulsar responsible for the cyclical activity of this complex object. (*Mount Wilson and Palomar Observatories*)

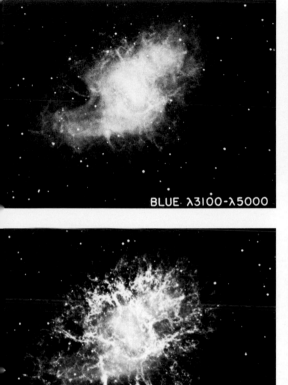

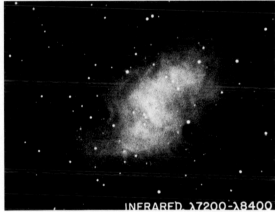

BLUE λ3100-λ5000

YELLOW λ5200-λ6600

RED λ6300-λ6750

INFRARED λ7200-λ8400

Four photographs of the Crab Nebula in Taurus. These are taken with the 100-inch Hooker Reflector at Mount Wilson and are photographed in blue, yellow, red, and infrared. (*Mount Wilson and Palomar Observatories*)

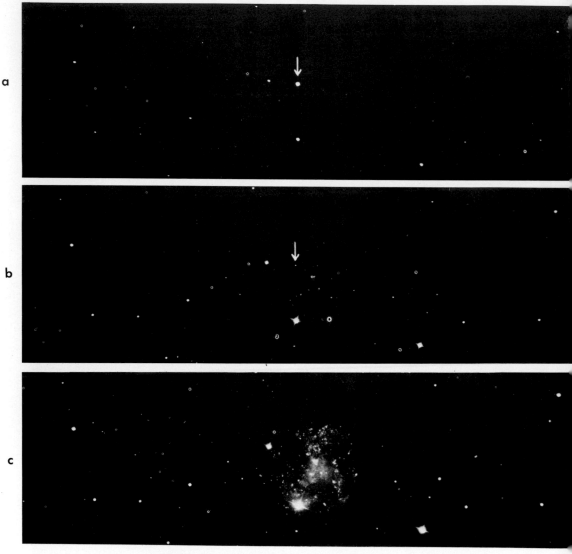

a) 1937 Aug. 23. Exposure 20^m. Maximum brightness.
b) 1938 Nov. 24. Exposure 45^m. Faint.
c) 1942 Jan. 19. Exposure 85^m. Too faint to observe.

Supernova in IC-4182. Note: This supernova was brighter than the entire galaxy in which it resides. A twenty-minute exposure on August 23, 1937, shows the supernova only with no evidence of a galaxy in this region. (*Mount Wilson and Palomar Observatories*)

Supernova in NGC-4725 in Coma Berenices.
Supernova is seen here on May 10, 1940; by
January 2, 1941, it had faded and disappeared. (*Mount Wilson and Palomar Observa-*
tories)

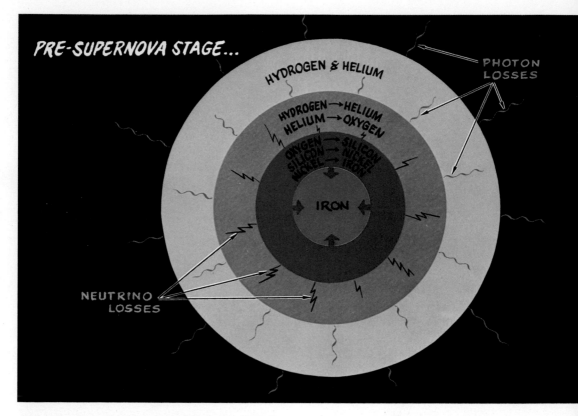

Section of a star in its pre-nova stage. The
iron core is surrounded by shells in which
hydrogen is converted to helium—helium to
oxygen—oxygen to silicon—silicon to nickel—
and nickel to iron. There are neutrinos escaping from the star at this stage, and photons of
various energies are being radiated.

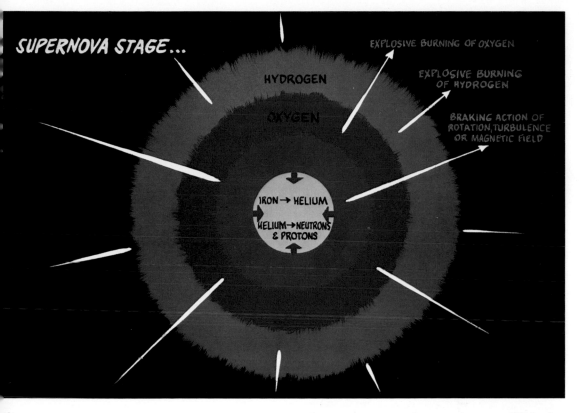

Supernova stage of a massive star. With the implosion of the core the iron decays into helium atoms and the helium atoms decay to neutrons and protons. Explosive burning of oxygen, hydrogen, and carbon occurs, which ejects a shell of matter away from the star at high velocities.

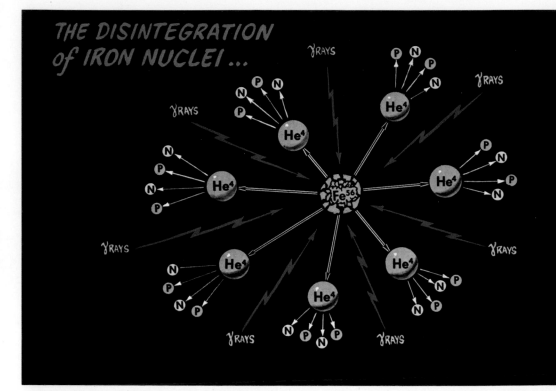

Disintegration of the iron nucleus. When subjected to a temperature of 5 billion degrees
the iron atom becomes unstable and fissions
into helium atoms which, in turn, disintegrate
into protons and neutrons. The energy absorbed from the star in fusing and building
up the elements to iron from protons must be
expended again when this process makes the
iron atom disintegrate. Only gravitational contraction can supply this inordinate but necessary amount of energy.

SYNCHROTRON RADIATION...

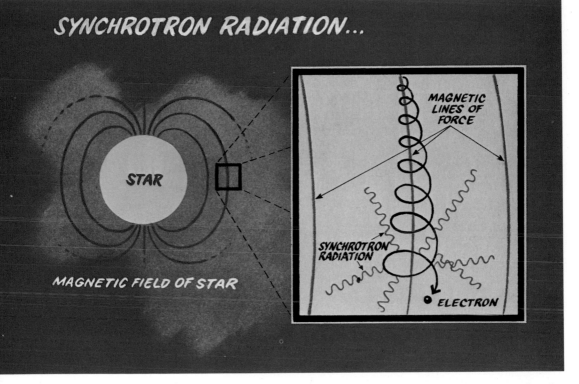

STAR

MAGNETIC FIELD OF STAR

MAGNETIC LINES OF FORCE

SYNCHROTRON RADIATION

ELECTRON

The motion of an electron moving under the influence of a magnetic field gives rise to radiation because the electron experiences a continual change in velocity and, therefore, radiates continuously. The intensity of the emitted radiation depends on the energy of the electron (a measure of its speed) and the strength of the magnetic field. This means that a wide variety of radiations can be emitted by this process—ranging from the high energies in X-rays to the relatively low energies in radio waves.

SUPERNOVA

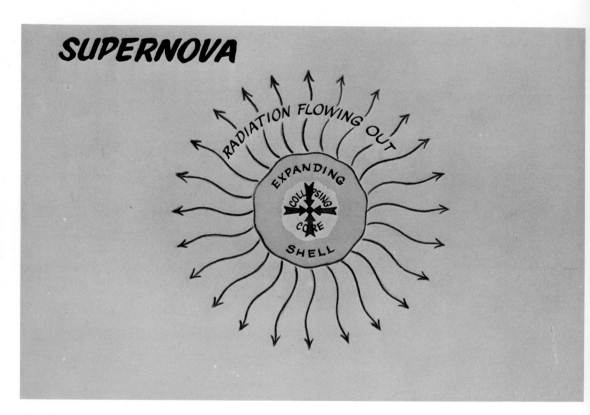

Conditions at the instant of the supernova eruption. The core has precipitously contracted into the tiny radius of a neutron star with a speed that leaves a shell of matter behind. Because of the high temperature the shell experiences explosive burning and begins expanding at high velocities. Enormous volumes of radiations, especially in the visible regions, are emitted in this early stage so that the star becomes billions of times brighter than in its pre-supernova state. (*This and the following three illustrations have been adapted from "Supernova Remnants" by Paul Gorenstein and Wallace Tucker, courtesy of* Scientific American.)

SUPERNOVA

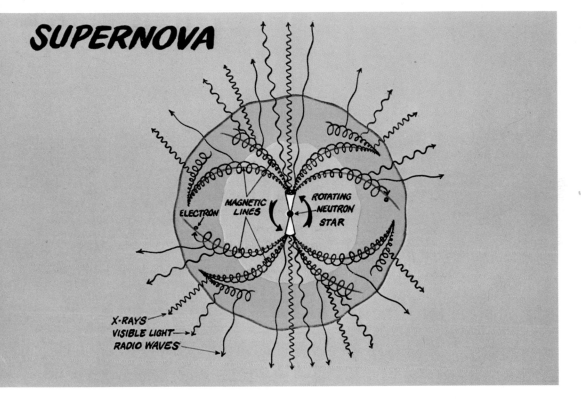

The swiftly rotating neutron star in the center possesses an intense magnetic field and produces a huge volume of electrons that travel at a considerable fraction of the speed of light. As the electrons spiral around the magnetic lines in the expanding shell they produce radiations of various types, ranging from low-energy radio waves to those of extremely high intensities in the X-ray region.

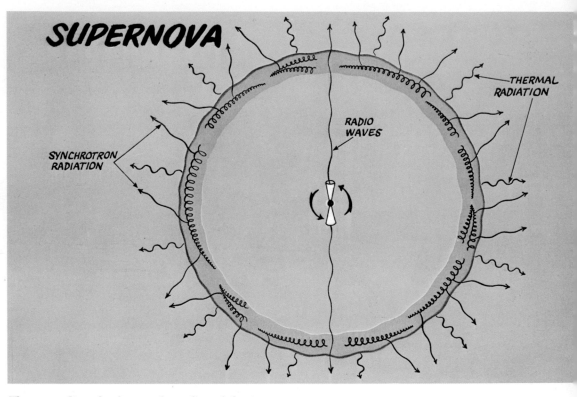

SUPERNOVA

THERMAL
RADIATION

RADIO
WAVES

SYNCHROTRON
RADIATION

The expanding shock wave has affected the interstellar medium to create a hot plasma, hot enough to generate thermal radiations. In the hot plasma, behind the shock wave, reactions take place which produce ultraviolet and X radiation. The high-speed electrons react with the magnetic field behind the shock wave to produce radio emissions. In this phase the rotating star is no longer the copious source of high-energy electrons that make the nebula surrounding the star visible. However, pulsed radio emissions from the central star still continue.

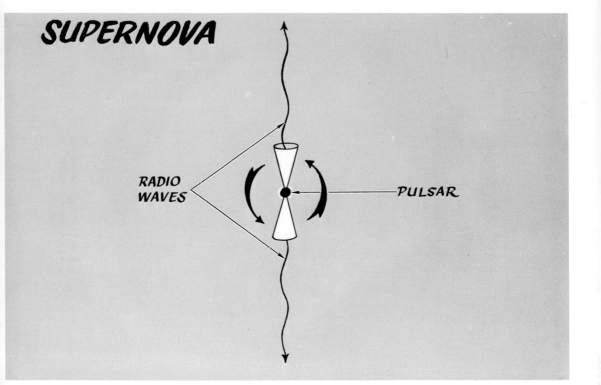

SUPERNOVA

RADIO WAVES

PULSAR

The energy emitted at the outset of the supernova explosion has been dissipated. The nebula has expanded until it is too tenuous to react with any form of radiation, or particle stimulus. But the central neutron star is still rotating very rapidly and is emitting pulsed radio emissions from special areas on the star. These emissions can be detected if one is in the proper plane to intercept them. The central star, because of these pulsed radiations, is called a pulsar.

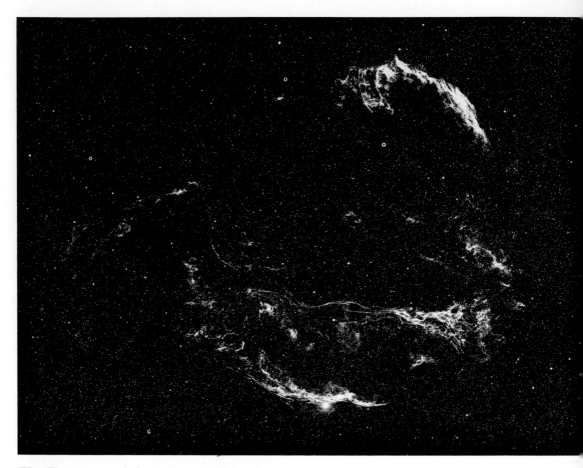

The filamentary nebula in Cygnus. This nebulosity resulted from an ancient supernova. Eventually, this material will cease glowing because the material will have been spread too thin to be excited by short-wave radiations. (*Mount Wilson and Palomar Observatories*)

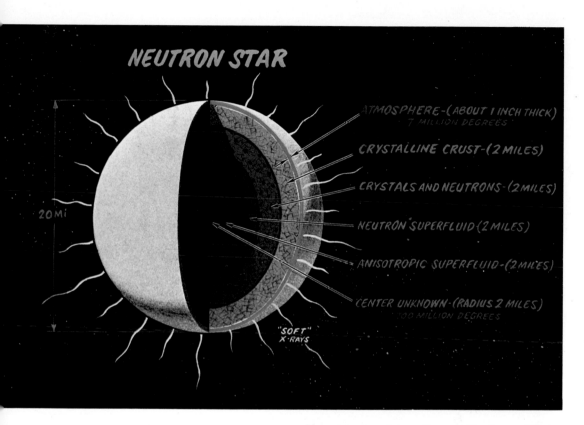

NEUTRON STAR

ATMOSPHERE-(ABOUT 1 INCH THICK)
7 MILLION DEGREES

CRYSTALLINE CRUST-(2 MILES)

CRYSTALS AND NEUTRONS-(2 MILES)

NEUTRON SUPERFLUID-(2 MILES)

ANISOTROPIC SUPERFLUID-(2 MILES)

CENTER UNKNOWN-(RADIUS 2 MILES)
100 MILLION DEGREES

20 Mi

"SOFT"
X-RAYS

Neutron star (according to Professor M. A. Ruderman). Section shows the minuscule atmosphere and the various layers as one penetrates to the center of the star. Soft X-rays are emitted by the star, which has a surface temperature of about 7 million degrees.

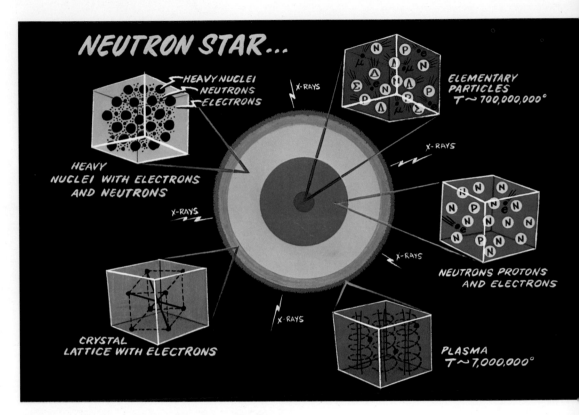

Neutron star (according to Dr. Remo Ruffino and C. E. Rhoads). Section shows the various layers, and the activities taking place in these various layers. The central temperature is on the order of 700 million degrees and only the most elementary particles are to be found in this central region.

many do move into the white-dwarf stage? Unfortunately, not all of the answers are as yet available.

One major step in this problem was resolved when astronomers pinpointed the positions of the central stars in planetary nebulae on the main-sequence diagram. Let's examine these celestial objects to understand the characteristics of their central stars.

On photographs, planetary nebulae appear as extended oval masses of gaseous material, with a faint, but very hot, star in the center. The oval masses are really complex, turbulent, concentric shell structures that are expanding with speeds of from 10 to 30 miles per second. Though they look like rings, they really are shells and in the shells the turbulent motions whip the gases around with speeds of up to 75 miles a second. In the few planetaries whose distances have been measured, it appears that the diameter of the shell is on the order of about 1 light-year—about 6 trillion miles. With the above expansion rate the shells will become too tenuous; that is, the gas will be too thinly strewn to be excited and seen after about 100,000 years. Most of the the planetaries we see have been generated within the past 50,000 years, with an age of 20,000 years being typical. The central stars are among the hottest known in nature. Their surface temperatures range from a cool 50,000 degrees to perhaps 1 million degrees. Because of this inordinately high temperature, most of the radiation of the star is in the far ultraviolet region of the electromagnetic spectrum. This ultraviolet is absorbed, digested, and reradiated by the gases in the shell to visible light which makes the shells visible. This means that the shells are considerably brighter than the central star which is the source of this energy, for the tremendous volume of radiation from the star is emitted in invisible form.

In analyzing the characteristics of the central stars, typical masses of these bodies appear to range from 0.6 to 1 solar mass. This is implied because greater masses are necessary to permit the heavier elements to be generated in the interiors. As has been seen, there can be no significant amount of hydrogen in the stars. However, the shells of gas are rich in hydrogen and helium and this fact leads one to question whether the ejection of this shell, which may amount to one-fifth the mass of the sun, is a necessary condition for the evolution of the hot central star. It is tempting to speculate that in the evolutionary sequence, stars are losing mass. Even the red giants have been shown to eject material and it is interesting to speculate that this type of mass ejection may be the most significant of all cases in which mass is lost. Some astronomers conclude that from 50 to 95 per cent of all the white dwarfs are *not* descendants of the planetaries. Thus, while a fraction of the white dwarfs are generically related to the planetary nebulae, at least half, or more, may have evolved from normal stars found on the main sequence and these do not pass through the planetary nebula stage. Unfortunately, the precise details by which they came into being are still shrouded by lack of information.

The entire picture of the formation of the white dwarfs is hazy and ill-defined. There are so many details lacking that, at best, the description of the evolutionary process becomes an exercise in conditioned speculation. However, the general conclusion is that many stars are losing some of their material and in this way are reducing their masses on their journeys to the final steps as white dwarfs eventually to be hidden in the celestial graveyard as black, invisible dwarfs.

6

Supernovae

About seven thousand years ago, in a remote corner of space, a star suddenly exploded and blew off a shell of matter. The star, which was relatively large and massive, had encountered critical internal-energy problems that threatened its physical integrity. When the limit was surpassed, it reacted in an indescribably spectacular and violent explosion—one of the most catastrophic in all the universe—to give rise to a supernova.

For six thousand years the light from the star in the constellation of Taurus, the Bull, winged its way outward. Finally, it reached the earth; the date was A.D. 1054. While science in Europe lay dormant and there was a hiatus in learning among the Arabs, observers half way around the world noticed the magnificently brilliant object in the early morning sky before sunrise. The light of that stellar cataclysm lit up the Eastern sky.

On July 4, 1054, Chinese astronomers looking into the early morning sky saw a brightly glowing celestial object that was much brighter than Venus. It was observed from Peking and Kaifeng and observers called this a "guest star." It was the brightest object in the sky, except for the sun, and was seen in broad daylight for 23 days, until July 27, 1054. Following its appearance in daylight, it began to dim but remained visible to the naked eye for 627 more days and finally disappeared from view on April 17, 1056. This bright star was independently seen by some Japanese astronomers who likewise recorded its appearance. Today, it develops that this was the brightest supernova on record, for it blazed with a brilliance equivalent to a half billion suns. If this star were at the distance of the nearest star, Alpha Centauri, we could easily read a newspaper by its light in the dead of night; it would have been considerably brighter than the full moon.

No record of this apparition is available from contemporary European sources but it must be remembered that this was the period of the Dark Ages when the torch of learning was all but extinguished on that continent.

There is a fascinating overtone to the discovery of this brilliant star. In 1955, William C. Miller and Helmut Abt, of the Mount Wilson and Palomar Observatories, discovered some prehistoric pictographs on a cave wall and also on the Navaho Canyon wall in Arizona. The picture, or petroglyph, on the canyon wall was cut into it while that on the cave wall was drawn with a piece of hematite, a red-iron ore. Both drawings show a crescent and a circle. Miller interprets the figures as a crescent moon and a star. He concludes that this could represent the supernova of 1054. There are two reasons for his conclusions: 1) that the moon was a crescent in the same position with respect to the supernova when it burst forth in 1054 and 2) that the potsherds collected at the sites indicate an Indian occupation at about that time. Thus, these pictures appear to be the artistic renderings of this supernova made by the early Pueblo Indians.

When the sky surrounding this supernova is photographed and the photograph examined carefully, it is found that the remains of the supernova now consist of a complex, chaotic, expanding shell of gas and contained in the gas are several stars. The entire complex of gas and stars is called the Crab Nebula. One of the central stars represents the source of all the material we now see as the nebula, for it is the star that exploded seven thousand years ago. This central star, as a neutron star, has a surface temperature of between 6 and 7 million degrees and an exceedingly small diameter. Today it is recognized as a pulsar—but more about pulsars later. In addition to photographs, the nebula has also been subjected to spectroscopic analysis. From photographs and spectrograms, the physical characteristics of the nebula can be determined.

Studies indicate there are two kinds of radiating regions in the Crab Nebula. The filamentary network is made up of gas streamers heated to some tens of thousands of degrees and ionized by the intense ultraviolet radiation of the central star. When analyzed, hydrogen, helium, oxygen, neon, and sulfur are found there. These filaments are seen projected against a background of a large, luminous amorphous region.

From photographic plates taken about a dozen years apart, it was discovered that some of the filaments in the nebula were moving outward. From the angular size, approximate distance, and expansion rate it was determined that if we went back in time about nine hundred years the nebula would have been concentrated in a point source. Thus, the Crab Nebula was directly linked to the supernova outburst seen by the Chinese and Japanese astronomers almost a thousand years ago.

Supernovae are not unique. There have been several well-authenticated apparitions in recorded history. The first is, of course, the Crab Nebula; the second was Tycho Brahe's supernova of 1572; and the third was Kepler's supernova of 1604. Recently, there has come to light a supernova in the constellation of Lupus, the Wolf. This result, from a study made by Bernard Goldstein, indicates the supernova flared up to become as bright as Venus at the end of April 1006. Nonthermal radio emissions from this remnant, the supernovae signature, were discovered in 1965. Astronomers calculate that each galaxy experiences a supernova once in from a hundred to three hundred years. As they watch other star systems—galaxies—in space, they see supernovae flare up occasionally to become brighter than the billions of stars which comprise the galaxy. Since 1885, when the supernova S Andromedae was discovered near the nucleus of the great galaxy in Andromeda, about one hundred and fifty have been discovered by astronomers. Of these, only three have been detected in our galaxy though there are many objects, such as the Cygnus Loop and the Cassiopeia A Loop, which are suspect and may have resulted from supernovae explosions in the Milky Way. The precise time for the Cygnus Loop is almost indeterminate, though astronomers indicate that if these are, indeed, remnants of supernovae explosions, the Cygnus Loop began its expansion about sixty thousand years ago. The Cassiopeia A Loop is the newest supernova in the sky, for this nebula began its expansion in A.D. 1700.

Lest we think that supernovae are a single species, like a black Chevrolet four-door sedan, they come in a variety of types. Some astronomers recognize five different types of supernovae though only two are usually considered the principal varieties, and these are simply called Type I and Type II. The difference between the two types involves luminosity, variation in luminosity, spectra, and number and location in particular regions, or types of galaxies. In the two principal types, the variations in luminosity with time are remarkably similar. In Type I there is a rapid increase in luminosity until it reaches a maximum or peak magnitude —19 (9 billion times the brightness of the sun) and declines in 20 to 30 days by a factor of about 10. After this initial drop, the luminosity fades slowly. This type is normally found in elliptical galaxies and only occasionally in spiral galaxies, which means they belong to the older generation of stars.

The Type II supernovae increase in brightness to a maximum of —17, about 6 times fainter than the Type I. After a slight initial decline in luminosity the decline is halted slightly and then there occurs a more rapid decline in luminosity than in Type I. Type II supernovae occur exclusively in spiral galaxies and, thus, they comprise a younger generation of stars which are not over 100 million years old. These, then, must be short-lived stars and from what we have learned they represent the more massive stars. Roughly 6 times as many Type II supernovae are observed compared to Type I.

The Type I spectra are generally featureless, which means very little can be learned from them. The Type II spectra are characterized by broad emission lines of hydrogen and helium due to the very rapid expansion rate of the nebula—up to 5000 miles per second. Astronomers are suspicious about the anomalies in the spectra, for they appear to be abnormally deficient in hydrogen, the most plentiful element in the universe.

Why does Nature produce these prodigies? How do they come into being? What is the mechanism by which a flareup occurs which rivals the brightness of scores of billions of stars? Where in the scheme of stellar evolution do they belong? What is the end product of the stellar explosion? These are but some of the questions posed by the astronomer when he observes these spectacular outbursts brighten a corner of the sky. For the answers to some of these questions, we must again explore the life history of the stars.

Dr. John A. Wheeler has pointed out that "It is one thing to analyze a star as nearly as static as the sun. Is it not quite another matter to forecast the fantastic dynamics of a supernova? We know how to predict in much detail the nuclear reactions that go on in the sun and the other stars, and the output of radiative energy from the surface. Can we speak with equal confidence about a star undergoing violent internal motion?" Can we from a fleeting glimpse of a star reconstruct its entire life history? He speaks about the first atomic bomb blast at Alamogordo, New Mexico, in 1945. This was a superb and dramatic example of man's understanding of how nature works, for this experiment was unbelievably different from anything in all past human experience and it was predicted correctly. Recently these mathematical processes have been used by scientists of rare competence to predict the hydrodynamics of a supernova. Thus, the dynamics of the supernova explosion lend themselves to a careful analysis with an interpretation that cannot stray too far from reality. Despite this, the cause of the supernova explosion remains a debatable hypothesis and the subject of considerable controversy. With this as prologue, let's look for answers.

Stars with masses exceeding the mass of the sun by about 20 per cent can in later life become unstable. The astronomer Subrahmanyan Chandrasekhar, in a brilliant theoretical exercise in the late 1930s, showed that when stars exceed the mass of the sun by about 20 per cent they tend to become unstable and sometime during the twilight of their lives will undergo catastrophic changes to bring the star into an equilibrium which will permit the star to die with dignity. Many astronomers have studied the last stages of stellar evolution and have explored the effects of large mass on these last stages. Universally they agree that when the stellar mass exceeds the "Chandrasekhar limit," awesome and violent changes are the order of the day.

We have seen that the stability of a star results from a critical balance between the gravitational forces of the star, which tends to contract it, and radiation pressure, which tends to expand it. We have also seen that in the last stages of stellar evolution, when nuclear fuels become exhausted, the pressure balance is maintained by the degeneracy effects. It is this degeneracy which can permit the star to become a white dwarf and spend the rest of its days in this configuration. The star can cool off as a white dwarf and end its life as a cold, lifeless, unseen celestial cinder in space.

When the mass of the star exceeds the Chandrasekhar limit, serious problems arise, for the degeneracy effect cannot maintain the pressure balance. The only recourse left to the star is to maintain high temperatures, because high temperatures represent the only alternative to maintaining the pressure balance. But, to acquire high temperatures, an energy source must be available in the star. In the normal evolutionary process, the star calls upon nuclear fuels to supply this energy. However, when the star arrives at the last stages of stellar evolution and the nuclear fuels that normally supply this energy are exhausted, what can the star do to acquire more energy? The star is not energy-bankrupt yet, for it is a large, massive object with a considerable fraction of its mass at large distances from the center so that it does possess gravitational energy. It is akin to a stone atop a high tower. By virtue of its height, the stone possesses potential energy. Similarly, the falling in of the outer layers of the star toward the center represents money in the bank, for it represents a vast storehouse of energy which is "on call." This means that the star, to maintain the pressure balance, can now begin to contract in order to supply energy.

How far will this shrinkage go? Does the star contract endlessly (we will explore this possibility in detail when we discuss the black holes) or is there some other mechanism whereby the star can eliminate the excess mass to achieve stability once more? Dr. Fred Hoyle and his colleagues have explored this situation exhaustively and have concluded that what actually takes place is a "catastrophic collapse, followed by a catastrophic explosion." It is the density generated by the collapse which triggers the explosion which eliminates the excess mass. Once the excess mass has been eliminated, then the star can once more pursue a normal course to extinction, for now degeneracy may take over to guide the destiny of the star.

The manner by which nuclear fuels burn and are exhausted in a step-by-step fashion is most intriguing and shows the application of knowledge derived from the laboratories to this process. Modern, ultra-fast computers have also played a dominant role in acquiring answers. Hoyle, and Dr. W. H. Fowler, in following through these steps, take as an example a star with a mass 3 times that of the sun and, thus, a good bit beyond the Chandrasekhar limit. A star of this mass would normally have a luminosity 60 times that of the sun and a lifetime of approximately 600 million years.

We have seen that the normal thermonuclear reactions occurring during most of the lifetime of a star provide the conversion of hydrogen into helium. When a significant fraction of the mass of the star is converted to helium, the temperature in the star's center rises and with this elevation in temperature—on the order of 200 million degrees—helium becomes a nuclear fuel and it is transformed into oxygen and neon. Thus, the helium core begins to generate a heavier core of these other two elements. The star now becomes a many-staged energy-generating system. In the thin helium shell surrounding the oxygen-neon core, energy is generated by helium being converted to oxygen and neon. In the thin shell, between the hydrogen and helium, the hydrogen is being converted to helium; again with the generation of energy. While this is going on, the core temperature is steadily rising. The shrinking of the star makes the core become more and more dense and raises the central temperature so that it increases from 200 to 300 million degrees. Even at this temperature, the oxygen and neon are quite stable and do not react. In time, the core becomes still denser and the temperature doubles, to 600 million degrees. Now the neon becomes a fuel, and magnesium and silicon are produced. The conversion to magnesium gives rise to free neutrons. When the star was born of the primeval materials, it contained some metals of the iron group. The free neutrons react with these metals to build up to the heavier metals, all the way to uranium—the densest of the natural elements.

Eventually, all the neon in the core is consumed. The core continues to contract and with this contraction comes a further rise in temperature. In the next step two atoms of oxygen are converted into a silicon and helium atom. Two silicon atoms combine to form an atom of nickel which then is transformed to one of iron. Not only are neutrons present, but protons and helium atoms are also available to enter into reactions to build up elements. Elements such as sulfur, aluminum, calcium, argon, phosphorus, chlorine, and potassium put in an appearance. Now the temperature of the core has risen to about 1.5 billion degrees. Again, the generation of heavier elements using free neutrons continues but a new factor enters at this stage because of the high temperatures.

Hoyle indicates that at about 1 billion degrees, powerful gamma rays are generated which possess the ability to disrupt the nuclei in the star. The neutrons and protons are stripped from the nuclei but this process is a self-regenerating one, for the protons and neutrons also recombine to provide a form of stability. As the temperature rises beyond 1.5 billion degrees, the stripping process becomes more frequent. One curious and unexpected result is that with the higher temperatures and the increased stripping and recombination processes, the nuclei accept more and more particles and, as a consequence, heavier elements are built up. At temperatures of from 2 to 5 billion degrees, elements such as titanium, vanadium, chromium, iron, cobalt, zinc, etc., are created. However, of all these elements, iron is the most abundant by a considerable margin. As before, with the conversion of light elements into heavy ones, energy is generated to keep the star from collapsing. The internal characteristics of the star now take on the aspects of an onion, with layers and layers of the various elements forming the principal constituents of the layers.

With the formation of the iron group, Hoyle points out that the star is set for its dramatic blowup. The nuclear reactions in the iron core give rise to a beta process involving an electron. This converts protons into neutrons. But, in this conversion, neutrinos are emitted and these now begin to carry off into space—away from the star—a considerable quantity of energy. At the high temperatures in the core of the star this energy loss is particularly serious, for it does not contribute to the radiation pressure necessary to maintain the equilibrium of the star. As a consequence, gravity is once more called upon to supply energy, and gravity responds by shrinking the star more and more and faster and faster to make good the major loss of energy through the emission of neutrinos. As before, with the shrinkage in the star's core, the temperature continues to rise and, finally, the core temperature reaches 4 to 5 billion degrees. With this temperature, a new factor is introduced. The core which comprised the iron group of elements undergoes a critical change and, instead of the iron group reacting to give rise to heavier elements, the iron is converted back to helium, plus a tremendous flux of neutrons. More neutrons are then available to be captured by the shell material to build up heavy elements.

At this stage, Hoyle indicates the star has reached a crisis. In the building up of the heavier elements energy was generated by the "fusion" process. Thus a vast amount of energy was expended over hundreds of millions of years. Now, the end product of these nuclear reactions breaks down into helium and, suddenly, the star is called upon to supply an energy quota equivalent to all the previously expended energy. But the star possesses only one asset—its gravity! To utilize this asset, the core must experience an extremely rapid rise in density, which means the core shrinks precipitously and, in a time on the order of one second, the "implosion" of the core takes place, withdrawing from the rest of the star. This is the beginning of the end for the massive star.

The implosion removes the pressure which had supported the outer parts, or shell, of the star. At the moment of implosion, the shell begins to fall inward to experience compression. The infall of the shell provides a tremendous amount of energy, for once more gravitational energy puts in an appearance. The result of this increase in energy is a further rapid rise in temperature and the infalling parts of the star are subjected to temperatures to which they are completely unaccustomed—temperatures on the order of 3 billion degrees. The light elements in the shell represent potential nuclear fuel for the star at 2.5 billion degrees. But to insure explosive burning, the temperature goes beyond this—to 3 billion degrees. Within one second, the kinetic energy of the infalling shell is converted to thermal energy and the shell substance is heated rapidly. At these high temperatures the lighter nuclei—principally oxygen —display explosive instability and begin reacting. As the density cannot exceed a couple of tons per cubic inch during this oxygen burning, it implies that the explosive burning takes place long before the infalling layers of the shell collapse onto the imploding core. It is estimated that the energy generated by these nuclear reactions in less than one second is equivalent to the sun's energy emission for a billion years! As the sun radiates 3.8×10^{33} ergs per second and there are about 3×10^{16} seconds in a billion years it means that the energy involved in a supernova outburst is on the order of 10^{50} ergs, that is, one hundred thousand billion billion billion billion billion ergs.

This awesome release of energy is sufficient to explode the outer parts away from the star, ejecting them with speeds of up to several thousand miles per second. These outer parts, composed of both the lighter elements and those of the iron group, contain a significant fraction of the mass of the star. Gases are blasted away from the star to create a nebula which extends over millions of millions of miles. The velocity with which it is endowed will continue to push this material away from the stars until after, perhaps, 100,000 years the nebular material is spread so thin (and is so diffused) that it no longer can be excited by the short-wave radiations of the intensely hot mother-star and, eventually, it fades from view. What is most significant is that magnetic fields are present—from the exploded material and from the interstellar gas. The compression of gas behind the shock-wave front compresses the lines of force, and the intensity of the interstellar magnetic field will rise to provide the energies to accelerate electrons. What is left is a super-hot star which has had its mass diminished so that it can now die and go to extinction with dignity. The star in all probability is a neutron star with a mass between 1.2 and twice the mass of the sun. The characteristics of the neutron stars we will explore in the next chapter. If the mass is in excess of twice that of the sun, then it might become a black hole, which, again, we will be exploring in a subsequent chapter.

7

Neutron Stars

When an iron poker is heated, it turns red. If heating is continued, the iron poker can become white hot. If still more heat is applied, the poker will begin to melt and, finally, by using a high-energy torch, the iron poker will become a gas and the iron atoms will literally explode away from the high-temperature mass. We are quite familiar with this process: everything on earth, if heated high enough, will eventually be transformed to a gaseous state. That is why the astronomer calls the stars incandescent spheres, for their surface temperatures are so high that the surface cannot remain liquid or solid. The surface must be a gas. This represents a veritable truth to us on earth. Astronomers have now discovered that Nature can raise densities to such high levels that gaseous materials can be frozen into a pattern to make the gas become a solid once more. It is not the kind of solid to which we are normally accustomed. Nevertheless, by terrestrial definition of solids, this high-density material must be considered a solid.

This sounds somewhat enigmatic and esoteric but the analytical probing and boring tools of the scientists indicate that this is the only way to explain the behavior of matter under certain conditions. The weird and the utterly ridiculous become reality in this fascinating realm of the neutron stars.

In the previous chapter we discovered that the more massive stars cannot end their lives in the normal fashion—by simply living out their lives and going to extinction over a long time span. In most of the massive stars the great bulk of material prevents this and the star, to regain its equilibrium, must blow off a shell of matter containing a sizable fraction of the star's original mass. Once the star has shed itself of its superfluous mass, it can then retire gracefully to old age but, to do so, it assumes a strange form. If the mass of the parent body starts off with twice the mass of the sun, the parent star may become a neutron star with some wholly formidable characteristics. Curiously, there are neutron stars with a mass one-fifth that of the sun but even in these cases the parent body was originally two or more times more massive than the sun.

In this discussion of neutron stars, it must be realized that the description of these physical characteristics is derived from theory and is highly speculative since the physical conditions in these bodies cannot be duplicated in terrestrial laboratories. What should be remembered is that this same analytical procedure has been employed with considerable success to predict the behavior and characteristics of other celestial objects. Thus, while the extrapolations (in most cases) are quite extreme and daring, there is some assurance that they bear some relation to reality. A scientist will, on occasion, test extrapolations by making a prediction. The prediction may turn out to be correct. If it is, then the scientist knows the extrapolation—no matter how wild—is correct and this may provide a point of departure for other speculations.

The concept of the neutron star is not new; the first speculation that such an object can exist was made by the imaginative Fritz Zwicky and Walter Baade in California in 1934. In the late 1930s it was also the subject of an investigation in this country by J. Robert Oppenheimer and G. M. Volkoff. The interest of these physicists was stimulated to attempt to determine the final state attained by a massive, collapsing star. Because the significance and, indeed, the implications of supernovae were discovered at about the same time, it was suggested that the neutron star may be formed in a supernova explosion and may be the remnant. Unfortunately, with the advent of World War II the focus of the science community was channeled into the war effort so that concentrated study was halted on these newly postulated and highly intriguing celestial objects. Following this hiatus, the study of neutron stars was resumed in the 1950s in a purely theoretical fashion to see where they fit into the problem of the creation of the elements in the centers of the stars. Thus, the neutron star holds the distinction of being the only major astrophysical object whose existence and properties were predicted long before the physical discovery.

In the early 1960s, the discovery of X-ray sources in the sky proved a major conceptual advance to scientists who began looking at neutron stars as a possible source of celestial X rays. Toward the end of 1967, a new and puzzling class of celestial objects was discovered in the sky—the pulsars. The discovery signaled a most significant event in the study of neutron stars, for it reopened the question of the origin of celestial X rays. A tremendous surge became evident in the study of these objects, for there were some aspects of neutron star behavior which were related to pulsars.

By a process of elimination, the only mechanism to account for the regularly pulsed radiation was a rotating neutron star and this once completely theoretical and speculative celestial body assumed reality and became subject to intensive research. No attempt will be made to discuss pulsars at this point—these will be thoroughly explored in another chapter. At this time, we will concentrate our discussion on the physical characteristics of neutron stars.

In neutron stars, gravitational forces are the dominant influence in their behavior. Following the explosion of the supernova, the parent star, which initiated the explosion by gravitational contraction, shrinks to a diameter of but a few miles. Various estimates of the diameters of neutron stars have been made and they range from about 5 to 100 miles. But, in this celestially insignificant sphere is packed the material of a celestial body like the sun, which is some 864,000 miles in diameter and about one-third of a million times as massive as the earth. The obvious effect of this concentration is that the neutron star is incredibly dense. In fact, it becomes so dense that it may become a solid star!

Now let's explore what goes on in the interior of these dense stars. What is the state of matter found on the surface, beneath the surface, and as one penetrates toward the center of these extraordinary dense stars?

Fully to define these physical characteristics requires that the astronomer know the "equation of state." This is a complex mathematical expression, describing the total physical state of a star. It takes into account such important properties as temperature, pressure, density, and rigidity and their intricate interplay. The only problem is that these properties are not too well known. Theoreticians indicate that if these properties were known, they could provide a full description of the star. But as these properties are not fully known, the physical characteristics of a star, in this case the neutron star, can vary widely, depending on what the particular astronomer selects as the equation of state. Let us consider two concepts of the neutron star.

We might begin by exploring the interior of a neutron star to see what type of particles are found in the center and what processes are associated with the production of the density characterizing such a star. To do this, we must again realize that a degenerate gas is present in neutron stars. As previously seen, electrons are not free to move in random directions. Were they to attempt this, the electrons would encroach upon regions occupied by other electrons which are already packed into the smallest possible volume. The electrons can only change speeds and locations as other electrons get out of their way. In a sense, the electrons behave like the crowd in a subway train during the peak rush hour. The passengers are all locked together and one can move only when others make way for him.

At the density of the central part, or core, of a white dwarf, the material consists of nuclei and the degenerate electron gas. However, as the star contracts beyond this point, the density increases and the electrons have to move more rapidly so that they do not violate the mandate that no two electrons with precisely the same energy can occupy the same place at the same time. Because their increased velocity endows them with a high energy content, the electrons can move into the nucleus of the atoms to react with protons to create neutrons. This step gives rise to special nuclei which can exist in equilibrium with the high-density matter but would be unstable and disintegrate in terrestrial matter. An increase in density further increases the speed of the electrons that are readily captured by nuclei, resulting in an increase in the neutron-proton ratio. When the electron velocities become sufficiently high, the neutrons which are bound in the nucleus now find that they can escape. The neutrons become "unbound" and move freely out of the nucleus. Following this, the neutrons themselves form a degenerate gas. The neutrons, like electrons, obey specific physical laws which preclude their encroachment on the domains of their neighbors. When a high enough density is reached, the "unbound" neutrons find themselves in equilibrium with neutrons in the nucleus. This would represent a stable condition were it not for the fact that there is an inevitable increase in the density. The end result of this increase is that the number of neutrons becomes perhaps a thousand times more numerous than the nuclei under equilibrium conditions. And still the density continues to increase!

With this further increase in density, a new situation arises. The protons that were locked in the nuclei begin moving out and eventually all the nuclei, as such, disappear, or, to be more precise, disintegrate. The remaining material possesses a very simple composition. It is simply a mixture of protons, neutrons, and electrons.

The protons and electrons represent 3 per cent of the mixture and the neutrons represent the other 97 per cent.

With increased density, another phase begins. Special particles, heavier than the proton or neutron, are generated. These particles—hyperons—put in an appearance when the density has increased to about three times the normal nuclear density. However, at this point we are dealing with too many unknowns and, as a consequence, no one can say with certainty what the core of the neutron star is like.

At these incredibly high densities, the degeneracy pressure of the neutrons and their mutual repulsion at short distances cause the neutron fluid to become virtually incompressible. Dr. Malvin A. Ruderman, of Columbia University, who pioneered in the study of collapsed objects, indicates that the star's "resistance to compression is 10,000 billion billion times that of ordinary steel." With a resistance of this magnitude, the star can supply the pressure needed to counterbalance the pull of gravity in the core of the neutron star.

In one concept, if a neutron star were cross-sectioned, it would present a relatively simple structure—but the complete explanation of what comprises the structure is virtually impossible. The neutron star will have a nondegenerate atmosphere, meaning that the top layers of atoms are similar to those we would find on the earth. If there were hydrogen or helium in the surface layers, these atoms would quickly diffuse into the interior, where, because of the high temperatures, they would be consumed rapidly in the thermonuclear reactions. It is considered highly probable that in the neutron star atmosphere the most abundant element is iron.

The density at the surface of the star is zero but, as we penetrate about a yard, the density increases sharply, to the point that a cubic inch would weigh 1500 tons. Surface gravity is so high that a man would weigh millions of tons! At the base of the atmosphere, electron degeneracy will set in and because of the high conductivity there will only be a slight increase in temperature as a person penetrates the surface. For a typical neutron star, the surface temperature could be expected to be about 1 per cent (or less) of the central temperature, depending on the presence of a magnetic field.

There is a fascinating overtone to the subject of intense gravitational fields on the surface of the neutron star. If one could imagine a man being set down vertically on the surface, the tidal force due to the difference in the gravitational attraction between the feet and head would pull the man apart by stretching him. Essentially, the feet would be pulled in faster than the head until the feet touched the surface. Standing on the surface, the man would be crushed to the thickness of the mucilage on a postage stamp.

The temperatures may be strongly modified by a magnetic field, for such a field affects the motions of electrons. If the star originally has a field, when it contracts to the dimensions of a neutron star, the magnetic field is concentrated and "frozen" into the star. It has been estimated that the magnetic field of a neutron star may have a value of 1 million million gauss compared to 1 gauss for the earth. While little is known about these magnetic fields, scientists have speculated that many of the cosmic rays accelerated in the Milky Way may come from the surface of the neutron stars, where particles have been accelerated by such powerful fields.

The inordinately intense magnetic field is not as uniform as the one we would find surrounding a spherical magnet. The supernova explosion represents such a violent cataclysm that the magnetic field associated with the exploding star would be knotted, twisted, and warped. This means that the relatively thin crust would contain an intense, distorted magnetic field which could subject it to tremendous stresses with the possibility of fractures occurring in the crust. Dr. Freeman J. Dyson, of the Institute for Advanced Study in Princeton, New Jersey, has suggested that the fracture of the crust could release superdense material from the subcrustal regions. One would then expect a "starquake" to begin and once this occurred it would result in a geyser or volcanic eruption to bring up core material to "load" the surface. The weight of this material could induce further fractures of the crust along with the release of still more core material. This process could continue in cyclical form to make the surface of the neutron star a violently turbulent area in constant agitation. Dr. Frank Drake, of Cornell University, indicates his observations suggest the presence of many tiny starquakes which may be reflected in pulsar periods.

A neutron star may be rotating very rapidly upon formation and if it slows down—and this has been postulated to account for the pulsar behavior—high stresses must develop in the crust. This could also create a fracture in the crust, resulting in a starquake which would again trigger volcanic and seismic activity. Thus, one may conceivably have a continual series of starquakes occurring during part of the life history of a neutron star. Indeed, some scientists suggest that the sudden decrease in the period of rotation of the Vela X pulsar may have possibly resulted from a starquake. However, these are conditioned speculations that appear to answer some questions but also raise others—which prove just as provocative and challenging.

Below the outer surface, for a depth of perhaps two miles, the material of the neutron star would increase in density by a factor of several million so that the cubic inch which weighed 1500 tons just below the surface would now weigh 4.5 billion tons. At the base of this layer would be found the neutron-rich heavy nuclei. These nuclei would be composed of elements varying from a relatively light weight, such as the molybdenum atom, to others more massive than any of the stable nuclei on earth. The crust would form a solid, for the nuclei would be arranged in a rigid crystalline lattice similar to some existing on earth.

There is a valid reason for the crust being a solid, and it deals with the form that an atom takes. The normal atom is surrounded by a cloud of electrons which screens it effectively from neighboring atoms. The interaction between atoms that give rise to crystals depends on the way the electron clouds interact. But, in a neutron star, we have seen that the electrons are free to move around in the internuclear regions. No electrons remain to screen the nuclei from one another and so the nuclei, being of like electrical characteristics, will repel one another. In this fashion, the nuclei arrange themselves in a regular lattice so that the nuclei stay as far away from one another as possible. This is why scientists believe that the nuclei in white-dwarf interiors and neutron star crusts will form a body-centered cubic lattice like iron, copper, or zinc but at higher densities. Can this crust be melted? The answer is "yes." However, the temperature to melt this crust would have to be 100 times higher than that normally associated with the surface of a neutron star. Curiously, the presence of the degenerate electron gas will provide a high conductivity for this crust. It will conduct electricity 100,000 times better than copper.

Dr. Ruderman indicates that in the least massive neutron stars, the crust may extend all the way to the center. This is only true for those stars with a mass 20 per cent or less than that of the sun. The more massive neutron stars have other layers as we penetrate deeper toward the center.

Going below the crust, one encounters a material consisting of a neutron-proton-electron fluid. This fluid is devoid of nuclei; the protons and electrons form a small percentage of the total fluid. It is expected that the neutron and proton fluids will form a superfluid. The neutron superfluid will probably possess properties similar to those of the helium isotope—helium 4—when it is cooled to near absolute zero. It might be remembered that helium 4 displays zero viscosity, which permits it to apparently defy gravity by climbing out of a glass. The proton superfluid, if one is formed, can be expected to be a superconductor capable of maintaining electric currents and magnetic fields indefinitely. The electrons in the neutron star interior moving at high speeds—within a tiny fraction of the speed of light—will also provide high conductivity.

Near the center of the neutron star, densities are so high that 1 cubic inch could weigh 15 billion tons, and the entire population of the earth could be squeezed into a pill the size of an aspirin tablet. The neutrons, protons, and electrons possess so much energy that they are converted into other exotic elementary particles which, on earth, would exist for less than a millionth of a second, but in the interior of the neutron star they are believed to be quite stable. Dr. Ruderman says, "The central core of a neutron star is a unique place in which one can expect to encounter phenomena totally outside our experience."

Another attempt to develop a physical description of a neutron star has been provided by Clifford E. Rhoads and Dr. Remo Ruffini of Princeton University. They base their concept on the equation of state worked out by Dr. Rolf Hagedorn of the CERN Laboratory in Geneva.

As in the first analysis, the neutron star is considered to be a many-layered affair, with each layer becoming denser as one approaches the core. In both instances, the physical parameters are similar. The radius of the neutron star is taken as 10 miles and a mass of 0.6 to 0.7 times the mass of the sun is assumed.

The outermost layer is the magnetosphere, consisting of a tenuous plasma of electrons and nuclei that are closely bound to the star's intense magnetic field. This is where originate the radio signals that serve as the signature of the pulsars. The ultrafast motion of the charged particles as they spiral around the magnetic lines of force give rise to various types of radiations. In some cases, the radiations will be in the radio part of the spectrum and in others they will be a short-wave radiation of high frequency. Just below the magnetosphere, the density reaches 15 tons per cubic inch, which corresponds to a density 100,000 times that of iron.

Below this outermost layer is the one with the characteristics of a metal. This is the "supersolid," with the material in this layer in crystalline form. The crystals are composed of atomic nuclei with atomic weights between 26 and 39 and between 58 and 133. The crystals are so small that one would have to line up 25 billion of them to cover an inch. The density in this layer is well over 1 million times higher than in the outer layer; to put it another way, its density is about 400 billion times that of iron.

On our way to the center, we drop into the third layer. This comprises a region of heavy nuclei, like cadmium, but is also rich in neutrons and electrons. The density of the third layer is 1000 times that of the second layer.

As we penetrate deeper into the neutron star, we reach the fourth layer, and the density goes up slightly—by a factor of about 5. However, at this density the heavy nuclei can no longer maintain their physical integrity and they break down into neutrons, protons, and electrons. Most of the material is in the neutron form. There are 8 neutrons for every proton and electron. The layer can be considered essentially a neutron fluid contaminated by electrons and protons. Below this layer lies the core of the neutron star. The density has increased by a factor of about 1.5 over the overlying layer. However, even this relatively small increase in density dictates that the particles move faster than in any other layer. The kinetic energy, that is, the energy of motion of the neutrons, along with the few protons and electrons, is so high that hard, or inelastic, collisions are constantly taking place. With this increase in density, the exceedingly fast-moving electrons are being transformed into negatively charged mu mesons and with a further increase in density the electrons are converted to negatively charged pi mesons. In this collision process all of the more than 1000 particles and resonances known to nuclear science are created and are to be found. In all probability, many more particles of which we have no knowledge are present under these conditions.

Whereas in the first model of the neutron star the core was solid with a resistance of compression 100,000 billion billion times that of steel, in this model the core is surprisingly soft. Dr. Hagedorn attributes the softness to the activity involved in the continual changing of the basic, exotic elementary particles into the various but short-lived forms. The forms change so rapidly that they simply could not exist for any significant time on the earth, if, indeed, they could be created.

The temperature of neutron stars is rather high. This is to be expected because of their mode of formation. It will be recalled that at the instant of the implosion which starts a supernova sequence, the central temperature reaches 10 billion degrees or more. But, in a relatively short time, in much less than a year, this temperature drops precipitously. The initial cooling proceeds predominantly by the emission of neutrino-antineutrino pairs. The emission of these particles soon drops the core temperature down to something on the order of hundreds of millions of degrees in the first 10,000 to 100,000 years. Following this phase, the star's temperature slowly decreases through radiation. The surface temperature also decreases rapidly. For a star with twice the mass of the sun, the surface temperature will drop about 2 degrees a year. While this appears inconsequential, it will drop from about 8 million to 6 million degrees in about 1 million years. In this cooling history of the star, we are ignoring the presence of a magnetic field which can accelerate the cooling by a considerable factor.

8

Black Holes and White Holes

I once saw a magician display a coin and then close his hand over it to make a tight fist. He invited someone from the audience—a foil—to come up and squeeze his fist hard. "Harder—harder," he said. The foil squeezed harder. Finally, when the fist was released, the magician opened his hand and the coin had disappeared. It had been squeezed out of existence, or so you were made to believe. Of course, the coin reappeared in the next few minutes when the magician skillfully plucked it out of the ear of the foil. This is one of a considerable number of tricks magicians employ in which solid objects are apparently squeezed out of existence only to reappear at a propitious time in some other strange setting.

Curiously, Nature also performs this trick: but instead of the small object apparently being squeezed out of existence, Nature squeezes out of existence a star several times as massive as the sun. Some scientists speculate that when the star is squeezed until it disappears, to an external observer it may reappear at some other time in some distant part of the universe or, perhaps, in some other universe. When Nature puts the "squeeze" on things, she is not kidding —she plays for keeps. To see how Nature performs her trick and its consequences, let's discuss the formation of a "black hole," one of those tiny, incomprehensibly dense stars from which neither matter nor light can escape. At this point, it should be stated that these same scientists believe a black hole may only be an intermediate step toward the completion of spherical gravitational collapse or a disappearance in what the scientist calls a "singularity." The singularity is a region of space-time where infinitely intense gravitational forces deform matter and photons beyond recognition and, as the radius of the sphere shrinks to a point of zero dimension and the volume goes to zero, matter and energy are squeezed out of existence. The singularity, to some scientists, is Nature's way of saying that the present physical laws we are using are not adequate to cope with the situation—perhaps because we have missed the proper application of some existing laws or, in the extreme, because new laws are needed. Other scientists are just as certain that once we have a black hole, the singularity is ruled out; they indicate that as it takes an infinite time to reach the gravitational radius and as the universe spans a finite time, the black hole simply does not have enough time to go to a singularity. But, more of this idea and its overtones later.

Perhaps an example will serve to illustrate what happens in space-time that could give rise to a singularity. Picture a thin sheet of rubber stretched over a large frame, and let us assume that this rubber represents a corner of the universe. If we take a ball and place it in the center of the sheet, the ball will sink into or depress the sheet to deform it. If we replace the ball with a heavier one and place it on the sheet, the ball will stretch the rubber more and the deformation will be greater, with the ball sinking deeper into the rubber. A still heavier ball will deform the sheet still more and the ball will sink still farther into the rubber. Finally, if the ball had an almost infinite weight and we assumed the rubber sheet could not tear, the ball would drop to an almost infinite distance from the frame supporting the rubber. And if at that instant the rubber sheet did open up a tiny hole, the ball might pop through the tiny hole to escape the sheet. With the escape of the ball, the pressure on the rubber sheet would be relaxed and it would spring back to its initial position as a flat sheet. The gravitational stress would have been removed from space-time, but the ball would have effectively left our universe. Where would the ball be now? This situation will be explored in depth.

The concept of a massive object "trapping" light so that it prevents light from escaping from it was first suggested by Pierre Simon de Laplace in 1795, when he showed that an object with the dimension of the orbit of the earth around the sun and with the density of the earth would be so massive that light could not escape from its surface. The object would be invisible to a distant observer.

To see how such an object comes into being we begin again with stars which have come almost to the end of their extended lifetimes and are now entering the final phase, when their active lives draw to a close. We have seen that in the case of normal stars whose masses are less than 1.2 times the mass of the sun, the star will eventually shrink to become a white dwarf and radiate away the remaining energy of the system over an inordinately long period of time, ultimately ending as a massive, dense, dark cinder, eternally wandering through the shoreless sea of space.

If the mass of the star is greater than 1.2 times the mass of the sun, then in some cases an unstable situation will arise; a substantial shell of stellar material will be exploded off in the creation of a supernova and the remnant body will become a neutron star of rather impressive characteristics. Finally, we deal with a star whose mass is in excess of two solar masses. This star may explode to give rise to a supernova; if the mass of the remaining material is still in excess of two solar masses, the star must condense into a tiny, dense body, for the gravitational forces completely overwhelm all internal pressures. Scientists believe that it is at this point that cataclysmic gravitational collapse sets in to give rise to the black hole. They further believe there is no stable state at the end of this star's thermonuclear reactions. There appears to be but one inescapable path open to the massive star and that is total and complete collapse until it has become an invisible black hole.

While this description appears an evasive answer to a most abstruse problem, let's examine the possible reality of this collapse in as much detail as is available to the particular mathematical expressions and framework we use in science today. We might begin with the historical background to frame it in its proper perspective.

In 1939, Dr. J. Robert Oppenheimer and a graduate student, Hartland S. Snyder, at the University of California at Berkeley, were studying the ultimate fate of a large mass of cold matter. It appeared that one of the more spectacular consequences of Einstein's general theory of relativity was that once this mass began collapsing, it could not stop and eventually would shrink to a black hole. Thus, once a large mass, such as a nonrotating, symmetrical star, begins to collapse under certain conditions, it will shrink down to a critical dimension known as the gravitational, or Schwarzschild, radius—named after Karl Schwarzschild, who first postulated this phenomenon. When it has reached this radius, nothing can prevent it from completing the collapse so that literally it closes down on itself. What is this gravitational radius? There is a rigorous mathematical expression for this which shows that for a body with the mass of the sun the gravitational radius is about 2 miles while for a system of over 1 billion stars—a galaxy—the gravitational radius would reach from the sun to the planet Uranus—about 2 billion miles.

Shortly after Oppenheimer and Snyder initiated their research, World War II erupted and this study was relegated to the background, not to be resumed until after the war. However, in 1963 a new class of objects suddenly appeared on the astronomical horizon. These were the quasi-stellar radio sources—quasars. The displacement of the spectral lines of these new objects indicated they were at enormous distances from the earth. Further, because of the tremendous red shift of their spectral lines, astronomers placed their distances at billions of light-years. For this reason, it appeared that they represent emissions of fantastically large amounts of energy. Without getting into particulars—we will describe them in great detail in the last chapter—one solution to the problem of this enormous surge of energy was to assume the gravitational collapse of a giant star or a cluster of stars. Astronomers believed that the only adequate source of energy to account for these colossal outbursts was gravitation. Thus, gravitational collapse was thrust into immediate prominence, and this idea surged forward with increasing momentum as scientists in many disciplines brought their ingenuity, skills, and resources to bear on the problem. Today, a decade later, scientists have acquired some understanding of the problem, and have begun to recognize some of the facets and shortcomings of previous theories which render the conclusions of their studies highly speculative.

What are the physical characteristics of the "black hole" and how do scientists hope to find them? Indeed, where should they look for these anomalous "things"? Many scientists have pondered these questions and some answers have become available to guide the astronomer in his search for them.

The very name—black holes—indicates that they represent a class of objects which cannot be seen. The gravitational field of these objects is so intense that if one could stand stand on the surface of a black hole and manipulate the most powerful searchlight from its surface, one could not see the searchlight even if one were no farther away from the black hole than the earth from the sun. In fact, if we could concentrate the entire light output of the sun into this powerful searchlight we still couldn't see it, for the light could not overcome the overwhelming gravitational field to escape from the surface. This is why the surface is called the "absolute event horizon." It represents the boundary of the black hole.

Scientists indicate that these strange objects cannot be understood within the framework of Newton's laws of gravitation. Gravity is so intense near the surface of a black hole that everyday Newtonian laws break down and must be replaced by Einstein's laws of general relativity. It is fascinating to see how these new laws apply.

One of the three astronomical consequences of Einstein's theory indicates that when light leaves a massive object, light is red-shifted because the light loses energy in overcoming the gravitational field to escape from the star. Radiation coming from a dense star—like the white dwarf accompanying Sirius A—has its light shifted toward the red. The denser the star, the more the light will be shifted redward, until there will be no light coming from the star in the visible region of the spectrum. But, if the gravitational pull of the star is increased by shrinking its size, then the force of gravity can become so overwhelming that light cannot escape from the star. Thus, it is completely impossible for an observer ever to see a black hole! Well, one might ask, if it cannot be seen, then how will we ever find one in the sky? To answer this question, the scientist resorts to clever stratagems.

Doctors Remo Ruffini and John A. Wheeler have studied this problem thoroughly and suggest several ways in which a black hole can be detected—but never directly. To begin, when a black hole is created by gravitational collapse, it should emit gravitational waves that could cross space with the speed of light and momentarily warp the geometry of space in which the earth is found. This warping would manifest itself by the gravitational waves simultaneously affecting and being registered on similar instrumentation in widely separated regions of the earth. The gravitational radiation could come from a star that was undergoing gravitational collapse. If this star, during its normal life, possessed any rotation, as it collapsed to become smaller and smaller, it would have to spin faster to conserve its angular momentum. Finally, a stage would be reached in which the velocity at the star's equator would approach the speed of light—the limiting velocity. At this point, the star would become deformed and might shed material. In this fashion, energy and momentum would be radiated away in the form of gravitational

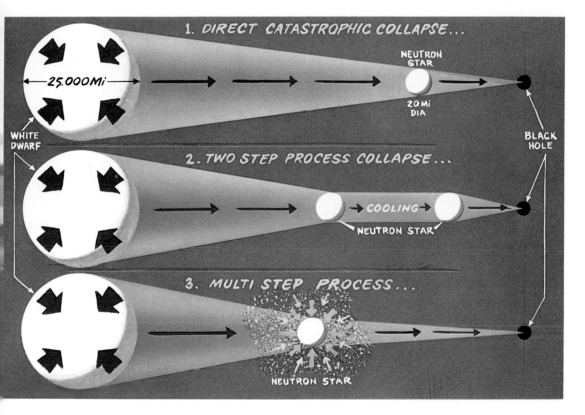

Steps showing the formation of a black hole. In each case, a white dwarf star is the starting point. 1) The white dwarf collapses to a neutron star and then to a black hole. 2) The white dwarf collapses to a neutron star which is too hot to shrink to a black hole. Thus, a cooling period must come into being before final collapse to the black hole. 3) The white dwarf collapses into a neutron star which has insufficient mass to continue to a black hole. The star accumulates material by accretion. Finally, when the mass is high enough, collapse to a black hole occurs.

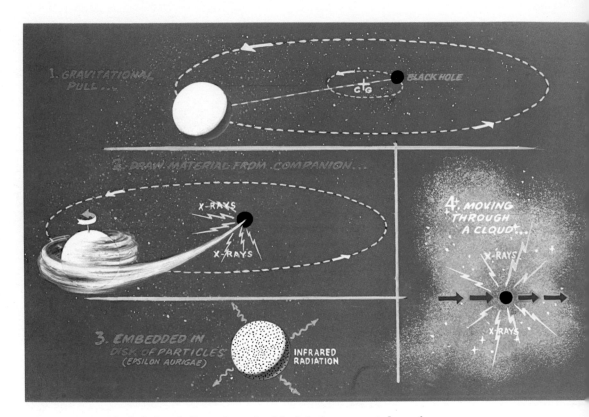

Detection of a black hole. 1) Detection of a black hole as a member of a binary system through gravitational interaction. The visible star will appear to oscillate around the center of gravity of the system without the second star being visible. 2) In a binary system, if the black hole can draw material to it from the companion, as the material enters the domain of the black hole, X-rays are emitted and they can be detected. 3) A black hole embedded in a disk of particles. In this case, excess infrared radiation coming from the system is indicative of the black hole around which the disk of particles revolves. 4) A black hole moving through a cloud of gas or particles will give rise to X-rays as the material is drawn into the black hole. (This mechanism is similar to 2.)

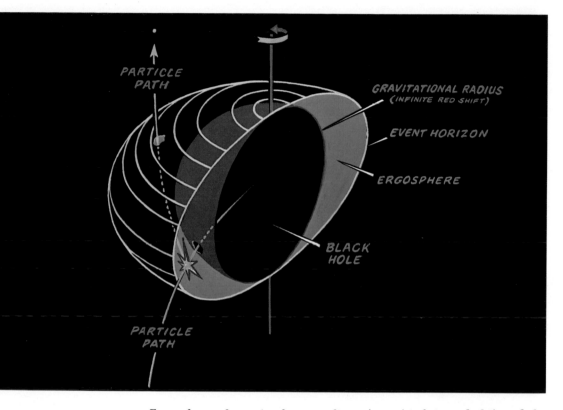

Ergosphere: the region between the surface of infinite red shift and the event horizon, where particles can still escape the black hole. A black hole represents the largest storehouse of energy in the universe. In the ergosphere a particle can enter and collide with a second particle orbiting the black hole. The particle that escapes has more energy, by half, than when it arrived at the ergosphere, while the second particle falling into the black hole has considerably less energy (one can think of this as negative energy). In this fashion 50 per cent of the rest mass of a particle can be extracted from the black hole and this efficiency is more than 50 times greater than the fission or fusion process of energy generation. It is, however, only half as efficient as the process by which energy is generated by the complete annihilation of a particle and an antiparticle. (*Courtesy of* Physics Today)

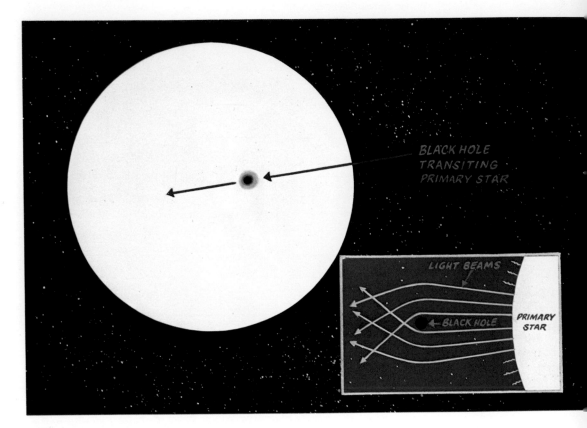

BLACK HOLE
TRANSITING
PRIMARY STAR

LIGHT BEAMS

← BLACK HOLE

PRIMARY
STAR

Black-hole transiting star in double-star system. If one had a super telescope which could observe a body under 10 miles in diameter crossing a small bright star, the black hole would appear to be dark gray in the center of its disk and then taper off to a very light gray at the edges. The gradation in shading is due to the gravitational bending of the light of the bright background star by the black hole into the region where the black hole should be observed.

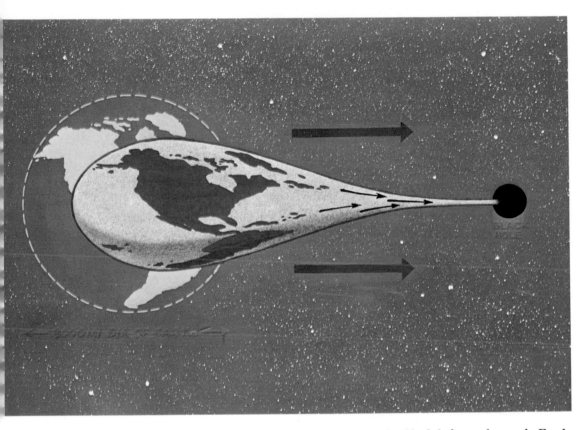

The action of a black hole on the earth. Earth approaching a black hole may be thought to become completely disintegrated, with the material "necking" down severely so that it would completely fall into the back hole. However, some scientists indicate this is not the way the earth would react.

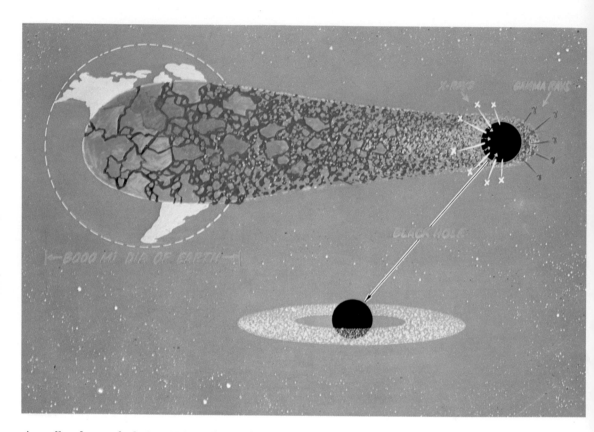

Actually, the earth, being 8000 miles in diameter, would be disintegrated and shattered by the black hole, which is but 4 miles in diameter. Some particles from the earth would enter the black hole, giving rise to X-rays and gamma rays. What remains would form a ring system spinning around the black hole, much like the rings spinning around Saturn. Additional radiation may be expected as the particles from the ring system spiral into the black hole.

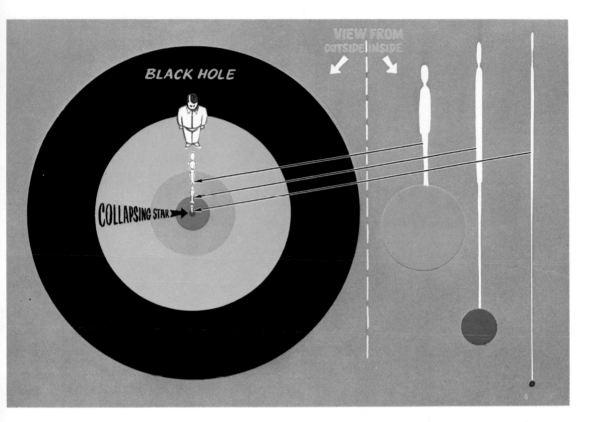

If an observer outside the gravitational radius possessed some mysterious powers by which he could see what was happening to someone riding the star down to its collapsed state, he would find that the man was thinning out because of the difference in gravitational attraction between the head and feet and, simultaneously, the tidal gravitational forces would compress him so that as he approached the gravitational radius he would be crushed to zero volume and have infinite length. However, someone riding the star down to its gravitational radius would not know that anything was different from his everyday world far from the black hole.

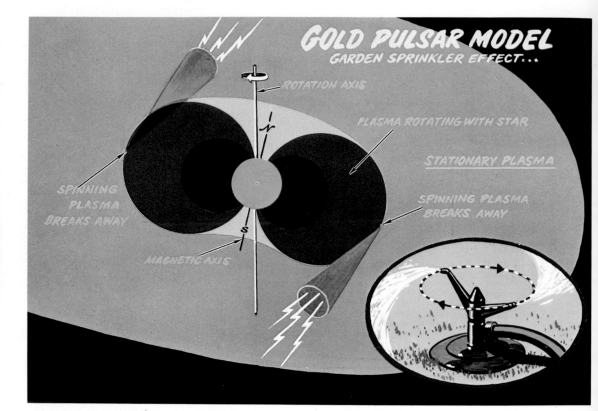

The Gold pulsar model. In this model of a pulsar, the magnetic axis is inclined to the pole of rotation. The plasma, rotating with the star, is disturbed (Gold attributes this to "sore spots"), and particles are ejected from these spots in a manner analogous to the way water breaks away from a rotating sprinkler. (*Adapted from "Pulsars" by Antony Hewish, courtesy of* Scientific American.)

The Ostriker model of a pulsar. A spinning star creates electromagnetic waves which can accelerate electrons to high energies. In this concept, the neutron star rotates rapidly and tends to wrap the lines of magnetic force around itself in a spiral-like configuration. An electrical field perpendicular to the magnetic field can accelerate particles away from the star—like riders on surfboards riding the crests of outward-moving waves. (*Adapted from "The Nature of Pulsars" by Jeremiah P. Ostriker, courtesy of* Scientific American.)

Light variation in the Crab Nebula pulsar. Using special optical instrumentation, the pulsar can be seen to brighten and dim optically in a period of approximately $\frac{1}{30}$ of a second. (*Kitt Peak National Observatory Photo*)

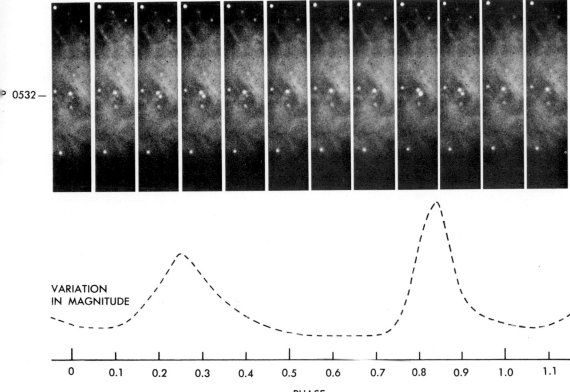

Light variation in the Crab Nebula pulsar. Using special optical instrumentation, the pulsar can be seen to brighten and dim optically in a period of approximately $\frac{1}{30}$ of a second. (*Kitt Peak National Observatory Photo*)

0532—

VARIATION
IN MAGNITUDE

0   0.1   0.2   0.3   0.4   0.5   0.6   0.7   0.8   0.9   1.0   1.1

PHASE

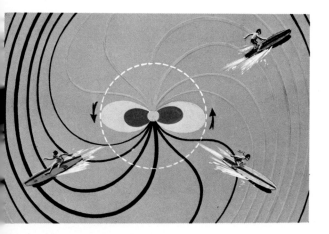

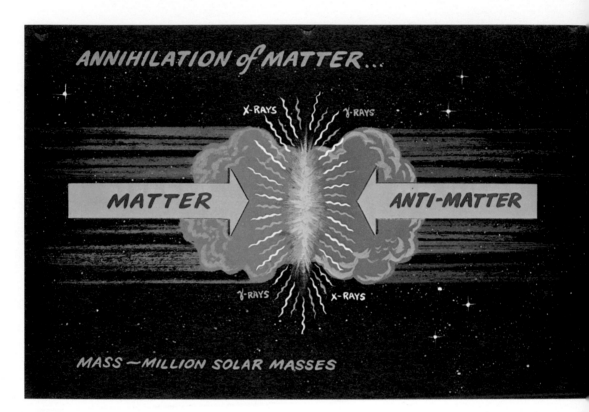

Annihilation of matter. If two volumes of material—one composed of matter and the second composed of anti-matter—could be brought together, the two would annihilate one another, converting the mass of the material to short-wave radiations. This is the most efficient energy-generating process known to science.

Quasar. It has been suggested that some galaxies collapse to create quasars. They would be spinning and as they became smaller the rotation would increase. As in the case of the neutron star, the magnetic field of the galaxy would be highly intensified as it became compressed.

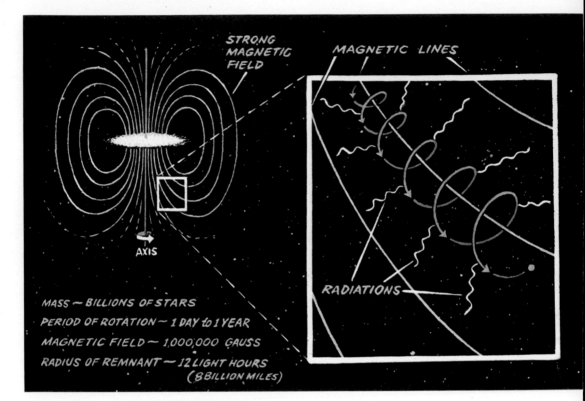

STRONG
MAGNETIC
FIELD

MAGNETIC LINES

RADIATIONS

AXIS

MASS — BILLIONS OF STARS
PERIOD OF ROTATION — 1 DAY to 1 YEAR
MAGNETIC FIELD — 1,000,000 GAUSS
RADIUS OF REMNANT — 12 LIGHT HOURS
(8 BILLION MILES)

Quasar. When the galaxy would reach the configuration of a stable star, it would be similar to a spinning neutron star. It would be spinning rapidly and have an intense magnetic field of perhaps 1 million gauss. The radius of the remnant would be about the size of the solar system but have a mass of billions of stars. The period of rotation could be from a day or less to more than a year. Synchrotron radiation could account for the quasar radiations.

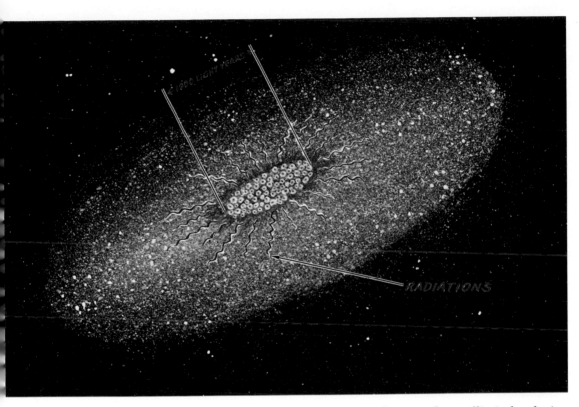

Quasar. In the amorphous elliptical galaxies the core is small and compact. Some scientists believe that these are massive stars which have brief life spans. When they reach the end of their evolutionary cycle, they are so massive that they become supernovae and because they all have approximately the same life spans, there occurs a chain reaction of supernovae, yielding the tremendous energy emitted by the quasar.

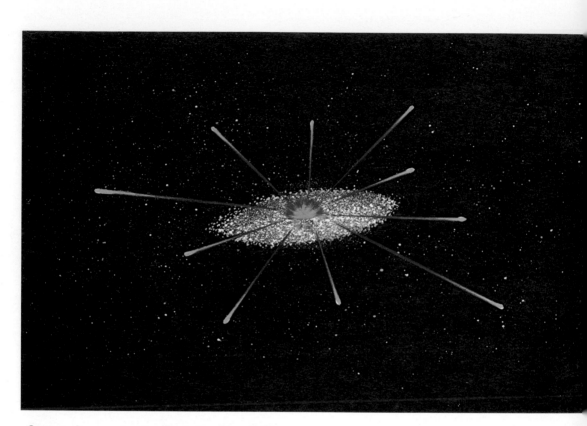

Quasar. Some scientists have speculated that a violent explosion occurred in the nucleus of a galaxy, giving rise to a tremendous release of energy. This energy was sufficient to blast away from the nucleus thousands of pieces of the galaxy with tremendously high speeds.

Quasar. Thousands of years after the initial explosion, the quasars are far from the galaxy and their segments are moving away from the galaxy, giving them all a red shift. The pieces are so constituted that they emit radiations in most regions of the electromagnetic spectrum.

A giant elliptical galaxy, considered the third most powerful X-ray source in the universe. This galaxy is also a strong radio source, known as Virgo A. Careful scrutiny indicates a luminous jet (some 5000 light-years long and strongly polarized) coming out of the nucleus. On other photographs a counterjet can be seen. (*Mount Wilson and Palomar Observatories*)

Quasar 3C-295. The arrow points to 3C-295— the brightest member of a remote cluster of galaxies in Boötes, photographed with the 200-inch Hale telescope. This is an intense radio source. An emission line in the spectrum of 3C-295 is indicated by the arrow. It is identified as a double oxygen line and these lines are red-shifted in the ratio of 1.47 to 1. The red shift places this quasar at a distance of 5 billion light-years. Because only one emission line appears in this spectrum, the identification is still speculative. (*Mount Wilson and Palomar Observatories*)

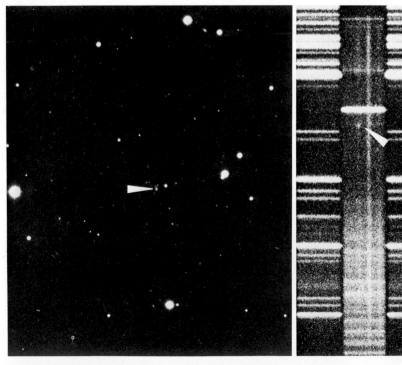

waves having frequencies in the thousand-cycle-per-second ranges.

At the Argonne National Laboratory outside Chicago and at the University of Maryland, Dr. Joseph Weber has set up his gravitational wave "traps." They consist of massive aluminum cylinders which should vibrate when gravitational waves encounter the earth. Weber's gravitational radiation detectors encompass both a high frequency (1660 cycles per second) and low frequency (1-cycle-per-hour) mode. The latter mode uses a sensitive gravimeter with the cross section of the earth as a sensor. The earth's natural quadripole frequency is 1 cycle per 54 minutes. This instrumentation should register simultaneous coincidences when gravitational waves come into the earth and, indeed, the traps are showing these coincidences. Unfortunately, the traps are catching too many—about one event per month—at this time. Perhaps in time, this strange situation will be corrected and pure gravitational waves may arrive and be detected to signal the gravitational collapse of a star. Some scientists point out that the results obtained by Dr. Weber, while extremely exciting with momentous overtones, have as yet not been verified. For this reason, many are skeptical about the detection of gravitational waves.

Dr. Weber appears to have pinpointed the direction of gravitational waves. He suggests that they may be coming from the center or nucleus of the Milky Way. However, it must be realized that the direction from which the gravitational waves appear to come is ambiguous. They can come from the galactic center or from the opposite direction, i.e., 180° from the nucleus. One finds this a most intriguing situation, for 180° from the galactic center we find the Crab Nebula, which, as we shall see in the next chapter, has a pulsar in its center.

One scientist skeptical of Dr. Weber's results is Dr. J. A. Tyson of the Bell Laboratories. He indicates that each event recorded by Dr. Weber would require for its generation a prodigious amount of energy. He estimates that this energy would be equivalent to the total conversion into energy of a body with the mass of the sun. There are no natural laws that preclude this. However, accompanying every conversion would arise a spectrum of radiations ranging from X rays to radio waves. These bursts of radiation have, as yet, not been detected.

Another method proposed by several Soviet scientists deals with the detection of broadband electromagnetic radiation emitted by matter as it falls toward the black hole. These scientists stress that the radiation is not emitted by the individual particles but by the entire cloud of gas when it is compressed and heated to 100 billion degrees by the funnel effect as it rushes toward the black hole and extinction. This effect is a manifestation of activity which lies just outside the gravitational radius and, thus, can escape from the black hole.

Scientists have devised still other schemes for detecting black holes. Dr. Kip S. Thorne indicates that unseen companions can be detected by the motion of some stars in the sky. This method has been used to detect unseen major planets (which are believed to be in orbit around several of the nearby stars). When this scheme is developed, it is found that the spectral lines coming from these stars are rhythmically shifted due to the "to-and-fro" motion of the visible star as it swings around the center of gravity of the double-star system. From the 800 or so stars which have been detected with the spectral-line swing, there are about 10 whose shifting lines indicate that the unseen star is much more massive than the visible one—somewhere between 1.4 and 25 times the mass of the sun. With masses on this order, it appears that the unseen stars must either be neutron stars or black holes. If they are neutron stars, we may find X rays being emitted by them. If they are black holes, we can only detect them through their secondary effects.

In the case of some close binary star system in which one star orbits the other, we might find one of the two is a normal star and the second a black hole. If this is the case, and material is being pulled out of the normal star into the black hole, the infalling matter should be heated by collisions to extremely high temperatures and, thus, they will emit X rays and gamma rays. These could be observed by X-ray and gamma-ray telescopes that are flown high above the atmosphere. However, in this case the problem encountered would be to differentiate the X rays coming from a neutron star and those coming from the proximity of a black hole. It may well be that if the suspect star is a neutron star, it will appear on a long-exposure photographic plate when taken with a large telescope as in the case of the Crab Nebula. If the second star were a black hole, it could never be photographed. Thus, in the case of close binary stars, with the aid of photography we may be able to differentiate between neutron stars and black holes to provide physical evidence of their existence.

In the event that the black hole is a member of an eclipsing binary, i.e., a pair of stars whose plane of revolution passes through the earth, why can't we see the black hole when it passes in front of the bright member of the pair? The answer is obvious. The black hole cannot be seen because of its small diameter. If a black hole passed in front of a star like the sun, which has a diameter of 864,000 miles, the light of the star would be dimmed by 2 billionths of 1 per cent. If we decide to be somewhat more realistic and have the black hole pass in front of a white dwarf, then the star would be dimmed by 2 millionths of 1 per cent. In either case, the loss of light is such that it cannot be detected with any current instrument. There is one intriguing overtone to this situation. If one could in some mysterious fashion see the black hole projected against the luminous star, it would not be a uniform black disk. It would have a black center and then shade off to a light gray at the edges. This curious appearance would be due to the intense gravitational field of the black hole bending the light of the bright background star.

In early 1971, evidence was presented by Dr. A. G. W. Cameron that there may be a black hole in the sky apparent through these secondary effects and he gave cogent arguments to support his hypothesis. He believes the secondary component of the remarkable binary star system, Epsilon Aurigae, is a black hole.

Epsilon Aurigae has been subjected to exhaustive investigations by astronomers in the past because of its enigmatic character. The primary star is a super-giant, with a mass 35 times that of the sun. The unseen secondary star has a mass 23 times the mass of the sun and should have a luminosity of 40 per cent of the primary—which it doesn't. The stars circle one another every 27.1 years and when the faint star passes in front of the primary, the eclipse lasts for about 700 days. The two stars are separated by 35 astronomical units, which is, very roughly, the sun-Pluto distance.

The two stars in the binary form a rather curious combination. The primary star can easily be seen and we have detailed knowledge of its physical characteristics. It is the secondary which represents the mysterious one, for it fits no known pattern of stars. Dr. Cameron indicates that this star is much too massive to be either a white dwarf or a neutron star. The only alternative is a black hole, which he calls a "collapsar." This he defines as a black hole created by a stellar implosion. From a study of the secondary, Dr. Cameron concludes that it is essentially a huge semitransparent disk of dust and gas revolving in orbit around an unseen central star. The semitransparent nature of the disk is called upon to reproduce the unusual effects noted during the eclipse. He notes that the mass of the disk can be determined and he sets it at no more than a few per cent of the mass of the central star. Because the central star has a mass of 23 suns, it means that the mass of gas and dust in the disk is about the mass of the sun. Some astronomers indicate they have detected excess infrared radiation coming from the Epsilon Aurigae system. If particles account for this suspected infrared radiation, then the source of the infrared must be a cloud of solid particles at a distance of 15 billion miles from the center of the star system, and they are assumed to be in the form of a large ring, or disk, at this distance. If the particles are tiny, the radiation pressure of the system will drive them away. Particles larger than this will spiral in, eventually to fall into the black hole. However, the characteristics of the system make it appear that the system so far is too young to have the particles spiral in. In time, it is believed they will be absorbed by the black hole.

What must be realized is that this system may represent the first visual evidence of the existence of the black hole. It is for this reason that astronomers may make the Epsilon Aurigae system an object of intensive study and scrutiny in the future.

Two other black holes have been postulated by Dr. Remo Ruffini and his Princeton colleague Dr. Robert W. Leach. They have indicated that the X-ray sources Cygnus X-1 and Cygnus X-3 may both be black holes emitting nonpulsating X rays. The pulsating X-ray sources are presumed to be neutron stars. Dr. Ruffini has further indicated that because X-ray double stars or binaries are so numerous, limits on the masses of neutron stars and black holes can be determined with a high degree of reliability. The Princeton team derived a mass for Cygnus X-1 of eight times the mass of the sun. Because of this mass and the knowledge that it does not pulsate, they strongly suggest this source to be a black hole. Cygnus X-3 is also a non-pulsating X-ray source with a 4.8 hour period of revolution of the two stars. Its distance lies somewhere between 25,000 and 35,000 light-years away from the earth. The binary system, because of its period, has to be extremely massive or the two stars would have to be close together, which means the system would have a short lifetime. If it had a short lifetime it is highly improbable that we would observe it. Thus the Princeton team concludes that Cygnus X-3 is a black hole. This gives us three candidates for black holes in our sky.

Let's get back to Epsilon Aurigae and assume that at some time in the distant future, an observer may get close enough to the secondary of Epsilon Aurigae to enable him to see what is going on there. However, we must also stipulate that he will be far enough away from the black hole so that gravitational attraction does not affect him. As we have discovered, this means that the observer must station himself outside of the gravitational radius. What will he see?

To begin, we must recognize the fact that he will not see the star; as we have discovered, neither particles nor radiation can escape from the star. So we will assume that some mysterious powers will be provided to the observer by which he can detect some of the physical characteristics of the black hole. Let us further assume that the observer just happened to be there when the star collapsed.

The observer would find the star contracting, and, as the contraction continued, the physical shrinking of the star would slow up and, curiously, the light emitted by the star would become redder, for the light would be reflecting the effects of the increased gravitational field. In effect, the collapse toward the gravitational radius would make the light—more specifically the photons—and the particles take a longer time to cover the distance from where they were first seen to the time they reached the gravitational radius. The reason why it would take a longer time involves the "time dilation" effect (in which time slows down in an intense gravitational field): the stronger the field, the more time is "stretched." This is why the collapse would appear to the outside observer to decelerate. When either the photons or particles enter the gravitational radius, they simply disappear. Only in the regions immediately outside of the gravitational radius could they be seen, and they would appear to dodge behind a curtain not to re-emerge. Dr. Thorne indicates that it would take a few seconds for a star with the mass of the sun to go from a normal star to a black hole; for a mass equal to that of a billion stars the process may take a few days. Thus, the observer would have to look sharply to see the change. However, Thorne also indicates that the "winking out" of the star may be hidden by the luminous outer layers that the star would have ejected before plunging within its gravitational radius. Because of the lack of radiation from the black hole, Thorne says, "The events that theoretical physicists predict happen there can never be proved." Well, almost never!

Now, let's imagine two observers: one on the collapsing star and the other far enough away to be unaffected by the collapse. Further, let us suppose that the observer on the collapsing star sends a series of uniformly spaced signals (either radio or light) to the distant observer to keep him informed as to what is happening. As the star gets closer to its gravitational radius, the signals that were sent at regularly spaced intervals, according to the observer on the star, would be received by the distant observer at more widely spaced intervals. If the observer on the star emitted a last signal just before reaching the critical radius, it would take an almost infinite amount of time to reach the distant observer, and if a signal were sent after reaching the critical radius, it would never be received because it would never leave the star.

This is the reason why the collapsing star appears to slow down as it approaches its gravitational radius. If there were a clock held by the observer on the collapsing star, the clock would appear to keep normal time, while to the distant observer the star clock would appear to slow down and at the gravitational radius, if he could see the clock, he would find it had stopped.

Let us assume there is another observer who decides to ride into the black hole with the photons. Now a unique set of conditions prevail. When the man stands on the surface of the star that collapses to its gravitational radius, he will approach the gravitational radius but in an infinite span of time. What happens to him?

Again, Dr. Thorne has explored this situation and concludes that the observer's body would experience different gravitational forces. He says: "His feet, which are on the surface of the star, are attracted toward the star's center by an infinitely mounting gravitational force; while his head which is farther away is accelerated downward by a somewhat smaller though ever rising force." This, we realize, is due to the height of the observer. Thorne continues: "The difference between the two accelerations (tidal force) mounts higher and higher as the collapse proceeds, finally becoming infinite (at the gravitational radius). The observer's body which cannot withstand such extreme forces is stretched between head and foot to infinite length as the radius drops to zero."

Simultaneously with this head-to-foot stretching, the observer is pulled by the gravitational field into regions of space-time with an ever decreasing circumferential area. This can be visualized if one thinks of an iris diaphragm on a camera lens. As the diaphragm is made smaller, the edge comes close to the center and the circumference of the diaphragm decreases. If the edge comes closer to the center, in the case of the star, the gravitational field will become stronger. As a consequence of the observer getting closer to the gravitational radius, tidal gravitational forces must compress the observer on all sides as they stretch him from head to foot. The circumferential compression is actually greater than the longitudinal stretching and, therefore, the observer, as he approaches the gravitational radius, is crushed to zero volume but has an infinite length.

Dr. Wheeler indicates that time and space in a black hole are interchanged in an unusual way. The direction of increasing time for the observer on the black hole is the direction of the decreasing radius. Thus, the observer has no more power to return toward the surface than he has the power to turn back the hands of the "clock of life" itself. He cannot even remain stationary; that is, where he is. The reason, as Dr. Wheeler indicates, is simple; no one has the power to stop the advance of time.

Up to this point we have been assuming that we are dealing with a nonrotating massive star that is perfectly spherical so that the solutions to the field equations are relatively simple. However, we must face reality and indicate that almost everything we see in the universe—stars, planets, some nebulae, and galaxies—rotates. Thus solutions that did not incorporate rotation were first-order solutions and less than complete. In 1963, Dr. Roy P. Kerr discovered that he could incorporate rotation into the solution of the field equations. One of the results that surfaced disclosed that a rotating black hole could indeed create a link connecting an infinite number of universes. He showed that while the black hole could contract further, it did not necessarily go to the singularity. As a point of fact there is a dramatic change in the character and properties of the singularity. This change sets the stage for a form of space travel in our universe or between universes.

Earlier we discussed the effect of a singularity on space-time and showed that it is possible that a black hole could literally escape from our universe by so severely warping space-time as to cut itself off from our universe. This intriguing picture of moving about our universe or into another universe resulted from the discovery that the Schwarz-schild solution really was two solutions. One of the solutions deals with the relationship between our universe and the black hole; the other indicates that the black hole can move on into another part of our universe or into another universe.

To return to our rubber sheet analogy, we can visualize a second rubber sheet directly under the first; as the black hole deforms the top sheet in some mysterious manner the bottom sheet is also deformed as a mirror image of the top. Or one can picture a softly inflated rubber balloon into which one is poking a finger. We will poke a finger in from the other side along a diameter. Now imagine a marble being pushed into the balloon by one of the fingers; the finger coming in from the other side of the balloon just touches it. Further imagine that the marble mysteriously passes through the two distended layers of rubber. When the pressure of the fingers is removed, the marble ends up at the other side of the balloon, diametrically opposite to the point where it was introduced. If we imagine the marble to be a black hole in this fashion, we have transferred it to another part of the universe. Or, if there were other balloons representing other universes, it is possible for the black hole to have moved to another universe. Thus our space traveler could use the contracting black hole as a means of moving through our universe in both time and space—even to another universe.

One serious drawback must be mentioned. At this time one cannot visualize an astronaut being compressed to the densities found within the black hole. However, this should not be considered an impossibility, for this concept possesses fascinating overtones. One must remember that if we move with the speed of light, time literally stops, dimensions in the direction of motion shrink to zero, and mass becomes infinite. One cannot help concluding that an astronaut traveling at the speed of light would have zero dimension with infinite mass. There is one difference—the singularity is a point while the astronaut becomes a line at the speed of light. Theory tells us that even though the astronaut was compressed to a line—this is what we on the outside world would see, if indeed it were possible to see him—the fast-moving astronaut would not notice any difference in shape, motion, or time. This exposition gives rise to a most intriguing thought. Perhaps at some future date, by moving just within the gravitational radius, an astronaut may be able to move to another universe.

While we have indicated that, if one passes through the gravitational radius, one moves inexorably into the singularity, one of the Kerr solutions indicates this is not necessarily so. But then we can rationalize the difference by saying that the astronaut cannot move with the speed of light, he can only approach it. Also the astronaut is not compelled to move into the singularity; he can skirt it.

We have left one question unanswered: what happens when the black hole moves to another part of the universe or to another universe? What is its appearance? Again, other perceptive scientists have applied their mathematical skills and talents to these questions.

Roger Penrose, professor of mathematics at Birkbeck College, University of London, has developed another dramatic consequence of this collapsing to a black hole. He also indicates that it is conceivable that the black hole will disappear and, eventually, make itself apparent at some other time in some other universe. He also states that the creation of a black hole during gravitational collapse is a critical indicator that something strange will happen to the geometry of space-time. His studies seem to show that a singularity will halt the collapse; by that he means, a black hole, or that the collapse will really continue to zero dimension and infinite density. The conditions for infinite density present a case for another universe attaching itself to ours at this point and, perhaps, the singularity will appear in this new universe. Or, it may even appear in some other place in our own universe. Because this is such a startling development, some astronomers look suspiciously at quasars and ask themselves whether the appearance of these enormously energetic phenomena are black holes from our or some other universe suddenly flashing into being as a "white hole."

It is conceivable that there is an intimate relationship between a black hole and a white hole. Dr. R. M. Kjellming, of the National Radio Astronomy Observatory, in Greenbank, West Virginia, believes it seems natural ". . . to suggest that black holes, which are singularities by which matter 'disappears' from the universe, are related in a genitive manner to 'white holes,' defined to be singularities from which matter and energy emerge, usually in the centers of quasars and galaxies." The reason why the word "seems" appears is the dominance of the conservation laws in physics. Kjellming indicates that ". . . at certain points, our universe is multiply-connected, or that two or more 'universes' are connected to each other through black hole–white hole singularities." It is through these singularities that the exchange of matter and energy can be effected.

Some scientists visualize the formation of a black hole as a small-scale model of what may eventually happen to the universe as predicted by Einstein's general theory of relativity. It is generally agreed that we live in an ever-expanding universe and one of the most significant and pressing questions in all of science deals with the nature of the universe, together with its past and future history. Certainly, all current observational evidence points to the expansion of the universe. However, one most provocative question today is: is this expansion rate slowing down, and, if so, will it in some scores of billions of years collapse on itself perhaps to give rise to a singularity similar to that we have just described? It appears that in the future we may derive some indication as to what path the universe is taking—but long before that we may be able to foretell the eventual fate of the universe by learning what transpires in the formation of a black hole and the physical laws which govern its behavior.

9

Pulsars

Early in 1968 a flurry of excitement swept the astronomical world. A new, intriguing class of celestial objects had been discovered which emitted precisely spaced periodic bursts of radio "noise." Because of the apparent constancy with which these bursts were recorded—the accuracy was 1 part in 100 million—it provided some excitement, for to some astronomers these could have represented precisely timed, intelligent signals being beamed toward the earth by what were facetiously termed "Little Green Men." The question posed was: are distant civilizations trying to communicate with us? When scientific activity simmered down to serious study of these radio sources, they were discovered to have some remarkably repeatable characteristics and variations which ruled against the concept of intelligent signals from deep space. But there was, indeed, a new, strange type of object in space—of which we possessed no previous knowledge.

We live in an era in which scientific discoveries can be verified almost immediately. Our communications systems are so rapid that within a few hours of the sighting of a strange object in the sky, there will be literally scores of telescopes, manned by highly skilled observers, trained in on it. Thus, these curiously varying radio signals instantly became the focus of intense study.

The origins of this discovery go back to August 1967, when Jocelyn Bell, a graduate student at Cambridge, England, was analyzing the records of certain small radio sources (made with the new,

large radio telescope at the Mullard Radio Astronomy Observatory).
The objective of the study was to determine the effect of the solar
wind or plasma (particle emissions from the sun) on celestial radio
sources. Scientists had speculated that the effect of this plasma was to
make the radio sources scintillate with a precisely repeatable
periodicity. Fortunately, one such radio source passed directly over-
head at midnight. When it was at that point it was in the shadow
of the earth and should have been free of the solar plasma effects,
for the solar plasma cannot be refracted or "bent" completely
around the earth into its shadow. When observations were made of
this radio source at its zenith, surprisingly it "twinkled." Because
this precision twinkling should not have been detected, it was subjected to further scrutiny. It was determined that the variation was
intrinsic to the source itself; the variation of the radio signals was not
due to any effect of their passage through space as they approached
the earth. With this discovery, a new class of objects—now called
pulsars—was acknowledged. Pulsar is a sort of mnemonic for stars
emitting "pulsed radio signals." Today about 60 pulsars are known.
What must be realized is that their discovery was an accidental by-
product of programed research—another classic example of seren-
dipity.

One of the most remarkable characteristics of pulsars is that
they emit bursts of radiation in almost every segment of the electromagnetic spectrum. In at least one pulsar, the Crab, these bursts
were found in the optical, X-ray, and radio regions, and proved to be
one of the strongest sources of X rays in the sky. The remarkable
property of the pulsars is that the periods between bursts are extremely short for celestial bodies as large as stars; they range from
0.033 to 3.75 seconds. This period is so short that originally the
pulses were thought to be terrestrial in origin. Their reality was
accepted only after scientists tried to explain them as a phenomenon
associated with a terrestrial source; this became much more difficult

than ascribing to them a celestial origin. However, with concentrated study and the discovery of pulsars in the constellation of Vela and the Crab Nebula, there was an overwhelming swing of opinion in favor of a rotating star.

Pertinent questions could now be posed: what kind of rotating star? How could a star generate a period shorter than a quarter second? One obvious answer was to search for pulsars in the sky, identify them, and then attempt to interpret them in light of known physical laws. While simple in concept, this became a formidable task, for a careful search of the sky was made with large telescopes that could reach down to magnitude 21 (1 million times fainter than the faintest stars visible to the naked eye under ideal conditions). But the results were negative! No star was revealed in this survey. Through clever stratagems, the astronomer could estimate the distance to several of these stars. The first ones were considered to be within 100 light-years of the earth, though a later one was determined to be 15,000 light-years distant. With some knowledge of the distance and radio intensity, an estimate could be made of the brightness of the radio sources compared to the sun; it was then determined that they were radio sources enormously brighter than the sun. When this became apparent, astronomers realized they were dealing with a new type of celestial body. The problem was to determine what this new object was.

What now begins to unfold relates very much like a detective story in which many clues are scattered about and someone has to examine each clue to try to fit it into some sort of pattern. Some of the clues will be obviously misleading—others will appear to provide some sort of solution at first glance but, when subjected to further scrutiny they will also be found wanting. However, enough clues have been explored so that the answer, while not obvious or definitive, does provide at least a tentative solution or a point of departure for further speculations.

If we examine celestial objects suspected to be pulsars, the list is short. We already have met the normal stars, the white dwarfs, and the neutron stars. These are apparently the only objects which could give rise to the short-period pulsations characteristic of pulsars. But, almost immediately, we can discard the normal stars and white dwarfs.

We know that the normal stars are quite large and have relatively low surface temperatures. These characteristics rule against the normal star being a pulsar. Under normal conditions, before a star can yield X rays, its surface temperature must be on the order of 7 million degrees. This is almost 1000 times higher than the temperature of a normal star. The second objection is the size of the normal star. In the case of the sun, with a diameter of 864,000 miles, it takes the light about 4.5 seconds to move from one edge to the other. As the observed pulse is only 20 milliseconds—20 thousandths of a second—it obviously cannot be a normal star; at the speed of light, in 20 milliseconds the pulse can travel across an object only 4000 miles in diameter compared to that of the sun. Thus, whatever star is generating the sharp pulses, it cannot be a normal one.

It is conceivable that the diameters of white dwarfs may be small enough so that the observed pulse can be emitted by this small object—but what makes it pulse? There are pulsating stars in the sky (such as the cepheid variables) and, perhaps, pulsars could be a form of pulsating white dwarfs. But, when the fundamental period of pulsation for a white dwarf is computed, it is found that the shortest frequency is about $\frac{1}{4}$ of a second. The pulsars have much shorter periods than this, which fact effectively eliminates the pulsating white dwarf. Astronomers have also investigated the pulsations due to a pair of stars orbiting around their common center of gravity. If the atmospheres of white dwarfs were just grazing as they revolved around one another, the minimum period of revolution would be 1.7 seconds—which is 50 times longer than the periods of the shortest recorded pulses. Another objection to this hypothesis is that if the two dense, white dwarfs were orbiting they would have to radiate strong gravitational waves that would rob the stars of energy and, as a consequence, the orbital periods of the two stars would decrease. This decrease would have been opposite to the increase noted in the pulsars under observations. Thus, white dwarfs provide no solution.

The last class of celestial bodies capable of short-period pulses are the rotating neutron stars. When the physical conditions are analyzed, it is discovered that the pulsation period for a neutron star would be on the order of $1/2000$ of a second; therefore, pulsating neutron stars could not be the objects, for the period is too short. If the parent star were a large star and it collapsed to this tiny size, it would have to conserve angular momentum, which means the small remaining star could rotate with a period of $1/100,000$ of a second. But, at this speed, centrifugal forces would rip apart the neutron star; astronomers thus believe that in the process of shrinking to a neutron star in some as yet inexplicable fashion, some of the angular momentum is dissipated and the speed of rotation is considerably reduced to less than 200 revolutions per second. Of course, we have seen that one way of dissipating energy would be the emission of strong gravitational waves, but this entire problem of gravitational waves is so hazy and obscure that quantitative answers cannot, as yet, be derived.

Dr. Jeremiah P. Ostriker, of Princeton University, indicates that energy can be dissipated by the emission of waves of magnetic-dipole radiation (we will discuss this later). Such waves, he believes, can carry off energy and angular momentum from a spinning star. From the rate of slowdown of the Crab pulsar, the hypothetical magnetic field of the star can be computed and this turns out to be 2.6×10^{12} gauss, which is what was indicated for a pulsar. Precisely what the mechanism is for this energy dissipation is uncertain but there exist two possibilities. The first is by the emission of gravitational waves, and we have discussed this. The second is by magnetic processes. What is known is that neutron stars are whirling, magnetic dervishes rotating at high speeds.

Now let's discuss another characteristic of neutron stars—intense magnetic fields. The magnetic field of a normal star can be as little as 1 or 2 gauss for the sun or several thousand gauss for a sunspot. The massive stars have magnetic fields ranging from a few hundred to 40,000 gauss. If a massive star decreases in size, the surface magnetic field will become more intense, for it varies as the inverse square law. By that we mean that if the diameter is reduced to half, the star's magnetic field will increase by a factor of 4. If the diameter shrinks by a factor of 4, the strength of the field increases by 16, and so on. Neutron stars are approximately 1/100,000 the diameter of the parent stars, and so the magnetic field of the neutron star will increase by 100,000 times 100,000 so that it has a strength on the order of 1 million million, or 10^{12}, gauss.

Following the shrinking, the net result is a tiny neutron star with a surface temperature of about 7 million degrees, rotating with a period of less than 1/50 of a second and possessing an extraordinarily intense magnetic field on its surface. The problem is to fit these characteristics into a neat, small package and try to explain the observed physical properties of pulsars. Although others had informally suggested that a rotating neutron star might be accelerating particles at the center of some supernovae remnants, Doctors Thomas Gold and Franco Pacini, of Cornell University, suggested that rotating neutron stars with intense magnetic fields could emit high-energy particles which could account for many of the characteristics of a pulsar. From this suggestion to the actual explanation is a long, arduous journey which has not yet been satisfactorily completed but the essentials are known and thus the general outline of the explanation can be detailed. To do this, we will again examine the Crab Nebula.

In November 1968, astronomers at the National Radio Astronomy Observatory in Green Bank, West Virginia, announced the discovery of two pulsed radio sources in the vicinity of the Crab Nebula. Later that month, astronomers at the Radio Observatory in Arecibo, Puerto Rico, identified one of these sources with a pulsar in the center of the Crab Nebula that had a period of about 1/30 of a second. What must be appreciated is that at that time the radio pulses were believed to be more precisely timed than the most accurate clock on earth.

Then, within the week, the Arecibo astronomers discovered that the period of the pulsar was lengthening at the rate of 38 billionths of a second a day. This would amount to about $1\frac{1}{4}$ seconds in a year. While this is an almost incredibly small rate of slowdown, it can easily be measured.

Astronomers know that if the period of this star is increasing, it is losing energy. From the period and characteristics of the pulsar, it was computed that if the pulsar was a rotating star it had to have the mass and radius of a neutron star; this meant it had to have the mass of about that of the sun and a diameter on the order of 15 miles. Thus, observation and theory indicated that the central star in the Crab Nebula was a pulsar, with the characteristics of a rotating neutron star.

Now we encounter one of those curious situations which, today, appears quite obvious but was never contemplated in the beginning of the pulsar study. Why had astronomers not found the active pulsar in the Crab Nebula before? The answer is that no one really knew what to look for. Once it was determined that it was a pulsar and radio signals were being emitted, both optical and radio telescopes could be turned loose on the Crab Nebula to scrutinize it and search for a common source of these signals. When this was done, the identity of the pulsar was established. Ingenious experimental astronomers assembled an instrument complex, consisting of a rotating slotted disk mounted in front of a television camera. It was essentially a stroboscopic image-amplifier. The disk was placed in the focal plane of the Lick Observatory's 120-inch telescope atop Mount Hamilton in California. Thus, the telescope worked as a stroboscope and managed to show the blinking star on photographs made of a Vidicon tube. This southwest component in the Crab Nebula is a faint star of the 15th magnitude (4000 times fainter than can be seen with the naked eye). In reality, the star is brightest at magnitude 15. It decreases in brightness to magnitude 18—a drop of 15 times in brightness—30 times a second. In this fashion, it was possible to construct a light-curve of the star's variation in brightness. The pulses coming to the earth consist of a primary, or main pulse and an interpulse of half the brightness of the main pulse. The radio pulses coming from the star were subjected to high-frequency analysis which showed that the energy did not build up and diminish gradually but the curves have spikes in them, showing that there was considerable structure in the radio pulses. The double-pulse profile also appeared in the X-ray analyses of the star's energy emissions. But the energy of the X-ray pulse was 100 times the energy in the optical pulse and 10,000 times that received in the radio pulse. This, then, is the behavior of the pulsar in the Crab Nebula. What occurs there to generate these pulses?

Many astronomers have studied this problem to seek an interpretation of what happens to the star to give rise to these vast outpourings of energy. Dr. Ostriker and Dr. James E. Gunn, his Princeton colleague, proposed one concept that is finding general acceptance, so we might follow their analysis of what occurs on the star. They postulate the presence of magnetic neutron stars spinning in a vacuum and emitting low-frequency electromagnetic radiations. To illustrate how this arises, they also assume that the axis of the magnetic field is oriented at an angle with the spin axis of the star. As the star spins, it creates an oscillating magnetic field that can radiate electromagnetic waves. These waves can accelerate electrons to very high energies. They indicate that the magnetic field close to the rotating neutron star will have a dipole field.

To visualize this, imagine a bar magnet laid on a sheet of paper, and sprinkle some iron filings on the paper. The iron filings will arrange themselves around the magnet so that at the bar end the lines of filings will be directly away from the magnet and at the edges they may curve to form a circle going from one end to the other. This is what is called a dipole field. A sphere has the same type of dipole field, but in three dimensions—like the earth. The lines of the earth's magnetic field move vertically from the magnetic poles— with some of them moving straight out and others curving round the earth to meet the magnetic lines coming from the other magnetic pole. And, in the case of the earth, too, the magnetic poles are not coincident with the north and south poles, that is, the poles of rotation. The geomagnetic poles are about 11.5° from the poles of rotation. The earth is rotating slowly; consequently, its rotation is not impressed to any extent on the magnetic line configuration.

In the case of a neutron star which is rotating at 30 times a second, the rotation does affect the lines of magnetic force and they tend to curve and, finally, far from the star, the lines of magnetic force wrap themselves around the star. As one can see with the iron filings around a bar magnet, the iron filings are tied to the magnet; move the magnet and you move the iron filings. In a similar fashion, when the neutron star rotates, the magnetic field rotates with it. But, there is a limit to the rotation. As one moves away from the star, the speed will increase and finally one will reach a point where the peripheral speed will equal the speed of light. Astronomers call this boundary the "cylinder of light." In the case of the pulsar in the Crab Nebula, the speed of rotation is so great that the distance from the center of the star to the speed-of-light cylinder is about 1000 miles! To visualize this, imagine the moon rotating at 30 times

per second. At this speed, the equator of the moon would be traveling at about 186,000 miles per second. As the speed-of-light cylinder
is approached, the magnetic lines will open up and will no longer
close on themselves. In other words, within the speed-of-light
cylinder the magnetic lines form closed loops; outside of it they
open up to spiral around the star.

As the magnetic lines of force move far from the star, the lines
begin to spiral out from the star. Thus, if we could take a neutron
star and shrink it in size so that it fit on a white page, and sprinkle
iron filings on the page, the iron filings would outline the inner and
outer fields. The inner field would move out very gradually away
from the star but in the outer regions the field would spiral away
from the star as it moved out. This magnetic field would also have
an electric field which would be radial or on a line directly away
from the center of the star. With these fields, Nature has a mechanism
for accelerating particles away from the star.

Astronomers are certain that charged particles are emitted by
the star. Certainly, this is true of the sun where the solar winds
represent an outpouring of charged particles. In a neutron star the
charged particles are probably ejected from the magnetic poles and
are then accelerated by the changing electric and magnetic fields.
Once the particles have been accelerated, they can radiate by a number of known mechanisms. Charged particles move more easily along
magnetic field lines than across them so that the magnetic lines
behave as a collimator, or channel, to produce a concentrated beam
of particles in the same way that the reflector concentrates the light
in a searchlight to provide a concentrated and collimated beam.
Thus, there arises a collimated beam of charged particles which can
provide radiations from the shortest X rays to the longest radio
waves.

Doctors Ostriker and Gunn have studied the action of charged particles emitted by a neutron star possessing an intense rotating magnetic field. Their mathematical solution indicates that the energy flow in the waves is perpendicular to the wave fronts. The particles absorb some of this energy and, as a consequence, will also be moving out perpendicularly from the star to the wave fronts. They are then moving with the outgoing waves; like surfers, they catch a piece of the wave and ride it out of the nebula.

As the wave fronts move farther and farther from the star, they become weaker and lose their grip on the particles which then become uncoupled from the wave. This method can account for the X rays coming from the Crab Nebula. Thus, the neutron star provides both the magnetic field and the particles as means of explaining the pulses which arrive at the earth from this star. Other scientists examining this explanation do not accept it, for they have difficulty in understanding how this mechanism leads to directivity and the variety of highly complex pulse shapes observed in different pulsars.

Is there any other way in which the pulses can be generated? Dr. Gold has suggested that the pulsar ejects charged particles (free electrons and protons in the form of a cloud of plasma) near the surface which would travel along the magnetic field lines until they reached the speed-of-light cylinder, where they would radiate synchrotron waves at high frequency. He further proposes that the plasma would leave the star's surface in only a few places—"sore spots." This could be considered vaguely analogous to sunspots. He requires that the radiating particles have velocities close to the speed of light and that they beam their electromagnetic radiation in a direction along the edge of the speed-of-light cylinder at the rotation speed of the neutron star in much the way water leaves a rotating garden hose. In this way, it would behave like a "searchlight beam" that sweeps the sky.

The width of the beam streaming from the Crab Nebula, if it is typical, is less than 12°. Thus for every 30 pulsars radiating in this fashion, statistically only one would have a beam intercepting the earth.

Dr. Frank Drake, who has studied pulsars at the Arecibo Radio Observatory, has discovered that the main burst of radiation in one of the searchlight beams is not uniform but consists of a series of such beams—a sort of shotgun effect. He envisages the searchlight beam as consisting of a series of smaller beams directed away from the radiation region but still contained within the cone outlining the searchlight beam.

The searchlight beam explanation accounts for the lack of pulsed radiation coming from all but 2 of the suspected 20 supernova remnants in the Milky Way. In order for the pulses to reach the earth, the earth must intercept the searchlight beam. If the beam goes above or below the earth, we will not know that the star is a pulsar. Only 4 of the 60 pulsars have these double pulses and it may be in these cases there are multiple spots to account for the 2 pulses and there may be two independent active regions responsible.

While many questions remain unanswered concerning pulsars and their emission mechanism, astronomers are looking beyond this immediate problem to view pulsars as significant investigative tools to learn more about the Milky Way and its citizens. The study of pulsars holds considerable promise for the further understanding of the evolution of the stars and the properties of highly condensed matter in magnetic fields. The pulsars may provide new clues to permit a definitive determination of the interstellar magnetic field. Study may also yield information about the clouds of charged particles in the interstellar regions and, perhaps, detail the fine structure of these plasma clouds. Finally, some scientists view some pulsars as a new precision type of astronomical clock with such great accuracy that it may be used for tests of the general theory of relativity.

Thus, a serendipitous by-product of an investigation on the scintillation of celestial objects has evolved into a new and fruitful avenue of research which promises to yield a clearer picture of a unique type of star and of our immediate surroundings in space.

10

Quasars

A glimpse into the far-distant past to view and contemplate objects which may have faded into oblivion long before the earth came into being, some 5 billion years ago, can now be made by astronomers photographing noisy radio sources in the sky. Following an extended debate concerning the distance to these radio sources, astronomers appear to have derived new proof that they are literally billions of light-years away and, thus, at "cosmological distances." When we view these faint objects on photographic plates, it must be remembered that we are observing some of them as they appeared some 5 to 8 billion years ago.

In this discovery, astronomers appear to have resolved a major astronomical dilemma. Dr. James E. Gunn, using the 200-inch Hale telescope, has discovered a quasar in the midst of an unnamed cluster of distant galaxies in the constellation of Perseus which pinpoints the quasar's distance at about 3 billion light-years. Thus, a controversy which has challenged the ingenuity, skills, and resourcefulness of astronomers and physicists since 1960 may be drawing to a close.

One of the significant overtones to this retreat into the remote past deals with the physical character of the universe at that time. The discovery has been made by Dr. Maarten Schmidt, of the Hale Observatories, that the intensity of the ionized helium line in quasars is so low (5 to 10 times less than in normal galaxies) that not only has the outer limit of the observable universe been thrust back in space and time but this outer limit appears to have a different astronomical character and composition from the space in our immediate vicinity.

Following World War II, radio telescopes came into general use and, in their observations of the sky, astronomers became conscious of intense radio sources in widely scattered regions. This was not surprising: the nebulous material which abounds in space is rich in electrons, and space is riddled with turbulent magnetic fields to provide the proper combination to yield radio signals. However, in 1960 a strong radio source was traced to a starlike object and it was this identification which triggered the study of quasars. It was called 3C-48, which stands for the forty-eighth object in the third catalogue compiled by the Cambridge astronomers in England. If this source had proven to be a star, it would, except for the sun, have represented the first radio star ever discovered. For this reason, astronomers became intrigued and excited.

Studies made of the number of quasars in the universe indicate that while there is a potential of about 15 million, many are so distant as to be invisible and others may have burned out. Today, it is believed that about 35,000 can be reached. This distribution means 1 quasar occupying an area approximately 4 times the size of the full moon. This further meant that there were plenty that could be reached with large telescopes.

To learn more about the physical characteristics of these exciting and intriguing starlike radio sources, large telescopes were trained on them and spectrograms were taken. The resulting spectrograms, tiny in size but pregnant with information, proved to be completely unlike a stellar spectrum. Upon careful examination, even what appeared to be normal star images on special photographic plates were discovered to be these curious objects. Multispectral photoelectric measurements were also used to identify them. Quasars were found to radiate strongly in the ultraviolet, which in itself constituted a subtle clue. This clue was used by Dr. Allan R.

Sandage, of the Hale Observatories, to search for them. He discovered that quasars can be distinguished from ordinary stars by comparing images on photographic plates sensitive to ultraviolet and ordinary white light. Quasars appear equally bright in ultraviolet and ordinary light, while other stars apparently emit much less ultraviolet radiation.

As astronomers were digesting this intelligence, another quasar was discovered in 1963. This was 3C-273. It was brighter than 3C-48 and when a spectrogram was taken of this quasar, several faint lines appeared and they occupied positions coinciding with those of a hydrogen series—if the series had been shifted toward the red end of the spectrum by 16 per cent. That the shifting of these spectral lines, proposed by Dr. Schmidt, could match the hydrogen series represented a tremendous stride in our understanding of quasars. With this clue, lines of ionized oxygen and magnesium were also identified and astronomers re-examined the lines from the quasar 3C-48. When these were remeasured, it was found that the lines were red shifted by 37 per cent. This was a considerable fraction of the speed of light, the limiting velocity in the universe. Since that identi-fication, literally hundreds of these sources have been discovered and the observed red shift varies from 15 per cent to over 340 per cent for the quasar OH-471, discovered in 1973. A red shift of 340 per cent indicates a velocity of recession of about 167,000 miles per second—90 per cent of the velocity of light! While this is in itself most exciting, when the explanation for this red shift is applied these dramatic objects appear to be scattered in space almost to the very edge of the observable universe. As an aside we might add that a velocity of recession of 167,000 miles per second corresponds to a distance of 11 billion light-years. Further, it is currently estimated that the "big bang" that hypothetically created the universe took place about 12 billion years ago. Thus the light from OH-471 started on its way only one billion years after "creation."

We have discovered that when lines in an object's spectrum are shifted to the red, it indicates that the object is moving away from us. (The greater the displacement of the lines, the faster the object is receding.) Thus, the shift of lines to the red in the spectrum of a celestial body means that it is receding and the velocity of recession is an index to its distance. In Chapter II, we discovered that there is a direct relationship between the red shift and the distance as determined by apparent diameter and apparent brightness, which enabled astronomers to measure the distances to some of the remote galaxies. This "Hubble law" remains valid—though it has been modified in terms of constants. Nonetheless, the velocity-distance relationship is a powerful celestial yardstick as no one has been able to disprove its validity. The "law" provides a method for determining extragalactic distances within 20 per cent of the true distance. While this is not as accurate as we would like, the red shift remains the sole method available to determine distances to the most remote galaxies.

When the spectra of many quasars were scrutinized, it was found that they were all red shifted. Not a single spectrum showed a blue shift. This again fitted in very neatly with the ideas of the astronomer; as they were all red shifted, it meant they were all receding and were believed to be partaking of the general expansion of the universe. This law, when applied to quasars, provided a spectacular jolt to the astronomers; they were not prepared for what was to come. And today, even knowing some of the answers, they still shake their heads in disbelief concerning these starlike objects. To visualize the profound difficulties associated with these objects let's discuss their physical characteristics.

When it was discovered that quasars were at cosmological distances, it was instantly realized that they must be enormously luminous. How else could an object be bright enough to be seen with a 6- or 8-inch telescope and still be at the distances where normally long exposures with the largest telescopes were needed to record their images on hypersensitive photographic plates? To convert these brightnesses into meaningful terms, the astronomer did some arithmetic and was astounded to find that some of these objects were emitting 100 times as much energy as the largest galaxies in the universe—those which contain in excess of 100 billion stars. The numbers derived for these energies are almost inconceivably high.

Let's examine a model of a quasar. The quasar emits energy at the rate of 10^{47} ergs per second, which represents 0.1 per cent of the energy emission of a highly energetic supernova. If we multiply this by the number of seconds in a year, we get about 10^{54} ergs. But some quasars have been in existence for several million years. If this is taken into account we find the energy output of the quasar is on the order of 10^{60} to 10^{61} ergs. A titanic explosion must have occurred to loose this flood of energy.

While this is an almost inconceivable amount of energy, there are no dictates in the laws of Nature, as known today, which preclude energy emissions of this magnitude. It became apparent that the astronomer had discovered a new class of super-bright objects at the fringes of the universe. What were they like?

To begin, they were very compact—starlike! Again, no law of Nature restricts the size of an object, provided, as we have seen, it does not contract within its gravitational radius. This meant they had to be enormously bright and small. The astronomer could obtain spectra of these objects and the spectra, aside from being red shifted, showed both emission and absorption lines. Astronomers know that emission lines are created by a glowing low-density gas. When the emission lines were measured, it was discovered that familiar elements, hydrogen, helium, carbon, oxygen, and neon were present. Further, these gases were ionized, for they had lost one or more electrons. Absorption lines were also found, indicating that a cool gas was absorbing some radiation coming from these objects.

There was some ambiguity associated with the absorption lines, for this cool gas could be at the quasar or it could reside anywhere in space between the galaxies. Actually, some quasars show absorption lines red shifted by varying amounts, which indicates these lines might arise from moving interstellar gas clouds along the line of sight between the emission source and the observer. Again, there is nothing in this description which is unacceptable to astronomers. Then came the problem!

When 3C-48 was first studied, there seemed to be a variation in the luminosity of this quasar. To verify this, astronomers examined old plates at the Harvard College Observatory. There was such a large margin of error associated with those plates that it was concluded there was no detectable variation. However, the first photoelectric study of this quasar (in 1963) showed a brightness variation of about 0.4 magnitudes over a period of about thirteen months. As 3C-273 was the brightest quasar—of magnitude 13 —astronomers began looking at other old plates (taken by the Harvard Observatory and by the Soviet Union's Pulkova Observatory), going back about 80 years. These plates showed significant light fluctuations, which occurred over a time scale of years with some "flashes" appearing with a much shorter duration—on the order of months or weeks. The most detailed optical studies were those made of 3C-345, which was observed to change in brightness by 50 per cent in a matter of twenty days, and one of the spectral lines, due to magnesium, was observed to split in the same period. Thus, 3C-466 turns out to be a spectacular quasar, for it brightened by a factor of 20 within a few months and in some cases its brightness changed by a factor of 2 in a few days. In addition to the optical variations, astronomers have also detected variations in the strength of the radio signals from quasars. For some quasars, this was about 40 per cent. There are even variations in the millimeter region of the spectrum. This continual variation of energy in various parts of the electromagnetic spectrum evolved as one of the most mystifying problems in trying to explain the behavior of quasars.

The time it takes a beam of light to move across an object is a measure of the diameter of the object. As an example, the earth is about 93 million miles from the sun. As light travels at about 186,000 miles per second, or about 11 million miles per minute, and if we divide 11 million into the 93 million miles of the earth-sun distance, it means that it takes light about $8\frac{1}{2}$ minutes to travel from the sun to the earth. If one should be on Pluto, then light from the sun takes about 6 hours to reach that planet. In looking at the sun, it takes light about $4\frac{1}{2}$ seconds to move across the solar diameter. It takes light 100,000 years to move from one edge of the Milky Way to the other. With this as background, let's project to quasars.

Some of these objects emit as much light as 100 galaxies of about 100 billion stars each. This means that the light source must be rather large for an emission of so prodigious an amount of energy. If this is the case, then how can there be fluctuations in the emissions of these objects over a period of days? It seems preposterous. If we moved Pluto out to about twice its distance, then it would take light about a day to travel from one edge of its orbit to the other. Thus, in the case of quasars which vary with a period of one day, the laws of physics dictate that the maximum size of the object must be on the order of a large solar system. And, if it is that small, how can we possibly pack 10,000 billion stars into it to emit its computed energy? It is this observation which provides the rock on which founders many theories proposed for the explanation of quasars.

One pertinent question arises: how do we reconcile the known data to derive a coherent picture as to what is taking place in space? One solution is to assume that perhaps the model placing quasars at cosmological distances is incorrect. Perhaps there is another acceptable explanation for the observed red shift—a gravitational explanation—but here we encounter even more insurmountable difficulties.

As we have seen, when a tremendous volume of energy is needed, the ultimate source of this energy is gravitational rather than nuclear. Because gravitation is capable of supplying about 100 times the energy per unit mass as nuclear, it is invariably called upon in difficult situations and the explanation of quasars presents just such a profound problem. What are the chances of the spectral lines being emitted in an intense gravitational field that can shift them to the red by such inordinate amounts?

To begin, we can postulate a body with the mass of the sun and concentrate this mass into a sphere a few miles in diameter. However, when we discussed the black hole, we found this was precisely what was needed to create the cataclysmic gravitational collapse of a star. Postulating an intense gravitational field is deficient on several counts. One objection is that it may take place so rapidly that we would never see quasars; the second is that the density of the objects, as determined by its emission lines, would be perhaps a billionth that required for the massive star. Another problem with the gravitational concept is that whatever results from gravitational collapse is a one-time affair. It does not recur. In the case of quasars, we see that they vary in brightness periodically. This means that gravitational collapse appears to be a blind alley from which no escape is possible.

If this explanation is not enough to doom the concept that the red shift is due to gravitational forces, we can go back to the emission spectrum. If the source is so massive, then under no circumstances could there exist the thin rarefied gas that, when excited, gives rise to the emission spectrum. Thus, on both counts, an intense gravitational field cannot supply the answer.

The only other mechanism involves recession. But, here, we have a further choice; a choice which does not require the objects to be so distant or so luminous. Dr. James Terrell had proposed that quasars are small objects which were exploded out of our galaxy, perhaps a half million years ago at a speed close to that of light, and have been moving away from us ever since. If this hypothesis is explored, one is faced with the problem of finding the energy to endow these massive objects with speeds in excess of 150,000 miles per second. An incredibly vast amount of energy is necessary to provide this type of ejection and propulsion.

Fred Hoyle has suggested that these objects may be at distances of tens of millions of light-years in place of the billions dictated by the cosmological red shift. There is visual evidence in photographs of galaxies which indicates that tremendous explosions are occurring in which substantial segments are moving away at high speeds and for long periods of time. But this motion is across our line of sight. If we assume that explosions in the galaxies gave rise to quasars, we should expect that half of the explosions would have taken place so that the material from these explosions would be moving toward us and show a blue shift. This shift of approach has never been observed. Further, if they are relatively near, their proper motion (that is, their movement across our line of sight) should be detectable. For example, an object at a distance of about 30 million light-years and moving with a speed of half that of light should show a motion of 5/1000 of a second of arc in a year which is detectable —and no such motion has yet been detected. To account for the absence of the blue shift, it has been suggested that there may have occurred a tremendous explosion in a nearby object and all of the fragments of the body involved in the explosion have passed the earth so that they are all receding and, thus, show a red shift.. However, if we assume this is the case, then there should be a preferred direction away from the earth in which those objects on the earth side of the explosion center should be moving slower than those on the opposite side. If this were the case, the distance would be most difficult to determine, for the velocity of the exploding objects would have two components. One would be due to the general expansion of the universe and the second would be the velocity due to the explosion. It is difficult to see how one would separate these two velocities to arrive at a realistic value of the distance.

A further objection to this hypothesis deals with the explosion itself. The initial explosion must have been inordinately powerful, since the material was endowed with high velocities. With a violent explosion and high velocities we might expect to find the material finely divided and uniformly scattered through space. But this is not the case. The matter appears compacted into many condensations. We know of no mechanism by which the compacting could occur, except gravitation, and gravitation would be overwhelmed by the initial high velocities.

One of the more significant problems dealing with quasars is the method by which they generate the energy-equivalent of 100 galaxies, each consisting of 100 billion stars. Never before has there been an occasion in which the scientist has had to propose theories to explain this prodigious volume of energy. For this reason, scientists have pondered the question and have proposed solutions, all of which are obviously highly speculative. Yet it is still rewarding to follow their thinking as they contemplate these highly involved and intricate questions.

Early in this exploratory search for energy sources, some scientists indicated that, with the supernovae representing the greatest release of energy in the shortest time, perhaps quasars evolve by the chain reaction of a cluster of supernovae in the nucleus of a galaxy— literally a sequenced cataract of explosions. But when computers were turned loose on the problem, it was discovered that (to derive the proper energy) a mass equivalent to about 100 billion stars was necessary to participate in this eruption. This result led others to look for a more powerful source of energy. Gravitational energy was suggested, for gravity could provide 100 times the energy release of a supernova. While this proposal is quite ingenious and can account for many details of the problem, there are too many unknowns to allow us to accept this concept as the solution to the energy-generation problem of quasars.

Another approach has been the investigation of massive super-stars that could undergo collapse to create this energy. In collapse, about half the "rest mass," or material of the object, can be converted to energy. There is, however, one known difficulty, which was discovered in our discussion of black holes. The mass cannot contract to the gravitational radius or the energy could not escape. The radiation is effectively isolated from the rest of the universe. Some modifications to this theory have been proposed in which negative energy fields are considered so that the super-star would oscillate in size about the gravitational radius to permit radiation to escape. Even if this were permitted, scientists do not know in what form this energy would be released. The massive-star model to explain the generation of energy appears attractive, but subtle observational features inherent in quasar behavior give rise to highly artificial concepts that are currently unacceptable.

Another attractive approach has been one in which matter and anti-matter coalesce, giving rise to the energy necessary to explain the observed properties of quasars. This method of energy generation is the most efficient in nature, for this represents all matter being transformed into radiation, such as gamma rays and neutrinos as well as electrons and positrons. For this concept to be valid, one must be sure that anti-matter is present in the universe and that it can remain separated from matter until it interacts in isolated cases. It has been shown that a mechanism of separation involving the use of a magnetic field is possible, but where does the anti-matter come from? No answer to this question can be found on the scientific horizon. Currently, this concept is not in vogue.

From what has been written, it becomes apparent that quasars represent a recently discovered phenomenon that has been carefully studied and detailed. Some of the physical characteristics can be explained, but the numerous questions posed concerning these objects indicate our basic lack of knowledge as to their origin, their character, and their life history.

To some astrophysicists the presence of quasars may provide a powerful tool to study matter in the intergalactic regions—that vast space between the galaxies. The detection of this thinly strewn matter is singularly important in the study of the size, shape, structure, and type of universe, for this study may reveal the total mass and density of matter in the universe. While we have some knowledge of the density of matter in the galactic regions, we are completely ignorant of what lies between the galaxies. Thus, the possibility of determining the density of this matter is highly intriguing to the cosmologist. In one of the quasars, the shift of the principal spectral line of hydrogen, called the Lyman alpha, is from the ultraviolet to the blue region which can be photographed from the surface of the earth. Scientists indicate that if quasars are at cosmological distances, then the passage of this radiation through intergalactic space and over these enormous distances would find some of the hydrogen atoms in space absorbing radiations in the spectral region between where the Lyman alpha line should be and where it is when photographed from earth. Thus, astronomers look for a general deficiency in the ultraviolet part of the spectrum. When the spectra of quasars were carefully scrutinized, this deficiency was deemed marginal and scientists concluded that only a tiny fraction of the intergalactic material was hydrogen. It was considered too low by a factor of about 100,000. To account for the lack of absorption, it was suggested that hydrogen might be present in an ionized form which cannot absorb this radiation.

Even from the absence of absorption, clues can be derived. If the gas is present as ionized hydrogen, there should be just as many electrons as ions present and they should markedly affect radio signals coming to us over cosmological distances. If free electrons are present, the radio waves of low frequency will travel at a slightly different speed than radio waves of higher frequency. If the distance is great, then the difference in travel time becomes significant. If the gas is present as neutral hydrogen, then quasars can be considered relatively nearby.

We began this chapter indicating that the work of Dr. Gunn —in identifying a quasar at the distance of the Perseus cluster of galaxies—was proof that quasars are at cosmological distances extending to the very edge of the universe 8 or more billion light-years from the earth. But there is some doubt as to whether this is the correct interpretation.

Dr. Halton C. Arp of the Hale Observatory has made an exhaustive study of "peculiar" double galaxies. Some of these galaxies appeared joined by bridges of stars and gas, and, thus, the two galaxies may be presumed to be at the same distance from us. However, when spectra are taken of these galaxies, it is discovered that the red shift is different for the two components! This is at variance with what is considered reality. If they are at the same distance, they must both show identical red shifts. But they don't. The only solution to this problem, in keeping with the validity of the red shift, is that they are widely separated but in the line of sight. However, only one or two cases of these double galaxies are known, and conclusions, based on this small sample, are extremely hazardous. If Dr. Arp's work is substantiated, then the keystone of our method of determining distance—our yardstick—fails and our studies of the universe will have suffered greatly.

What I have attempted is to present the various theories to account for what is seen and explain the mechanisms generating the optical phenomena. While all possess indequacies which preclude a definitive answer, it appears that the solution involving cosmological distances is currently the most plausible. Accompanying this solution is the problem of how to account for the incredibly large amounts of energy at these enormous distances. Astronomers and their colleagues in other scientific disciplines will have to rework their concepts and theories to account for these objects.

Even if the definitive solution to this problem is not realized in the immediate future, it should be recognized and patently obvious that the discovery of these strange celestial objects led to a revolution in scientific thought that dealt with the evolutionary process on the scale of the entire universe. If the cosmological approach is carefully studied and if this study comes to fruition, then science may be on the verge of a major breakthrough that will intimately link the smallest particles in the universe with the largest. One of the answers which may evolve from these studies will be the understanding of explosions whose magnitudes just a few short years ago were unheard of and unimagined. The next few years may be most fruitful in terms of our knowledge of the most distant skymarks in the universe.

Epilogue

And so we have come to the end of our story dealing with some of the paradoxes in this universe, for those that have been detailed do not represent the sum total. For every one we have laboriously brought into focus there may be dozens which lie submerged, awaiting the searching probe of the perceptive scientist. The author fully realizes he has dealt with all too few of these curious but more discernible objects that reside just beyond the known universe. Scientists are gradually becoming aware of the many more that will intrigue and fascinate them and compel them to delve deeper into that almost forbidden realm. From what we have learned, we further realize there are many more paradoxes on the astronomical horizon whose study could reveal the operation of new, unknown laws of nature which may endow many of our current theories with a welcome support or a disheartening obsolescence. It appears inevitable from what has been discussed that we have made dramatic additions to the corpus of scientific knowledge.

Today we are entering a new and exciting era. We have witnessed the unprecedented development of astronomical instruments which have broadened our vistas. We have witnessed the dawning of the space age, bringing with it a new potential to permit us to observe radiations from astronomical phenomena by rocketing high above the obscuring veil of the atmosphere. We have seen instrument-carrying rockets and small astronomical observatories orbiting the earth to reveal hitherto unknown celestial bodies. Indeed some scientists have indicated that one of these instruments—the X-ray telescope—may eventually have a more telling impact on astronomy than the 200-inch Hale telescope at Mount Palomar. These X-ray telescopes have revealed X-ray sources which appear to abound in the sky. However, as yet the astronomer is at a loss to explain the features of these objects. His knowledge of these sources is tantalizing in its incompleteness and represents a synthesis of observations, theoretical probings, and speculations all directed to the search for definitive answers.

The X-ray telescopes are but a single element in the armament-arium of the scientist. He looks for additional sophisticated instru-mentation to scan the sky and perhaps lead him to profoundly important breakthroughs in his explanation of the behavior of the objects he sees. Make no mistake, the radiations from the sky affect all of us, for the X-ray, ultraviolet, and infrared radiations penetrate deep into our bodies and deeper into our minds, even if they do not register on that evanescent film we call the retina.

Frankly, the scientist is mystified by the plethora of these para-doxes which flash meteor-like across his horizon. He sees exploding galaxies and calculates that energy emitted by these mammoth ex-plosions dwarfs to insignificance what is emitted in the most violent supernova explosion. He sees a string of galaxies which appear re-lated, as though members of a family, and finds that although they are not at the same distance from the earth, yet they appear to be con-nected by gigantic filaments of gas. The scientist hypothesizes celestial objects with the mass of a million suns apparently playing key roles in the activity he observes in the nuclei of these galaxies. He tries to find a role for the Seyfert galaxies in his scheme of the universe, but with minimal success. He speculates that the nucleus of a galaxy could contain relatively few supermassive stars, each with a mass 100 million times that of the sun. This, apparently, is the only conclusion he can derive when he attempts to explain the presence of enormously strong radio sources. The scientist ponders the role of neutrinos in so-lar and stellar evolution. How can he account for the incredibly vio-lent activity in the nuclei of exploding galaxies when we compute that the energy output we observe is hundreds of times that emitted by the

Milky Way system with its 100 billion stars? And to complicate the picture still further, this extravagant source of energy appears to be restricted to a tiny area of the nucleus. By what mechanism does a module-filled jet surge out of the nucleus of a galaxy, and what gives rise to the counter jet?

This is a brief listing of some of the paradoxes present in our universe, without an explanation of their presence or their place in the scheme of nature, for unfortunately we cannot subject them to the white illumination of rational experiment. The descriptions and explanations found in this book are in part conditioned speculations, and the scenario written may therefore be alien to the truth, for already we find some of them aesthetically unacceptable. But nevertheless they describe the objects we see. These are the objects that excite us. It is for this reason that scientists from every major country on earth have observed them, have pondered them, and have tried to provide some rational explanation for their being. Even with this concentration of effort, we feel it is only barely possible that the explanations presented have some relation to reality, for the definitive descriptions have not been developed.

What has been discussed in this book appears to be a relatively small sample of what exists in the sky, of which, in some cases, we catch the barest premonition. We can only speculate on the nature of these objects, which appear to trespass beyond the bounds of the known universe. And indeed there are many objects we see that even defy speculation. But speculate we must, despite the fact that what may evolve is irrelevant. If the scientist does not speculate, if he does not observe, if he does not experiment, and if he never contemplates—then he will never know.

Index